AUG 2 3 2022

CBD

Raphael Mechoulam in a lecture circa 1964 explaining the structure of cannabidiol in comparison with that of $\Delta^9$-Tetrahydrocannabinol (on the blackboard behind him).

# CBD

What Does the Science Say?

Linda A. Parker, Erin M. Rock, and Raphael Mechoulam

**The MIT Press**
Cambridge, Massachusetts | London, England

© 2022 Massachusetts Institute of Technology

This work is subject to a Creative Commons CC BY-NC-ND license.

Subject to such license, all rights are reserved.

The MIT Press would like to thank the anonymous peer reviewers who provided comments on drafts of this book. The generous work of academic experts is essential for establishing the authority and quality of our publications. We acknowledge with gratitude the contributions of these otherwise uncredited readers.

This book was set in Stone Serif and Stone Sans by Westchester Publishing Services. Printed and bound in the United States of America.

Library of Congress Cataloging-in-Publication Data

Names: Parker, Linda (Linda A.) author. | Rock, Erin M., author. | Mechoulam, Raphael, author.
Title: CBD : what does the science say? / Linda A. Parker, Erin M. Rock, and Raphael Mechoulam.
Description: Cambridge, Massachusetts : The MIT Press, [2022] | Includes bibliographical references and index.
Identifiers: LCCN 2021035141 | ISBN 9780262544054 (paperback)
Subjects: MESH: Cannabidiol—pharmacology | Cannabidiol—therapeutic use | Analgesics—therapeutic use | Neuroprotective Agents—therapeutic use | Anticonvulsants—therapeutic use
Classification: LCC RM666.C266 | NLM QV 95 | DDC 615.7/827--dc23
LC record available at https://lccn.loc.gov/2021035141

10  9  8  7  6  5  4  3  2  1

# Contents

Preface vii

1 Introduction 1
2 Chemical and Pharmacological Aspects 11
3 CBD/THC Interactions 39
4 Epilepsy 53
5 Neuroprotection, Tissue Protection, and Cancer 63
6 Nausea, Vomiting, and Appetite 97
7 Pain and Inflammation 117
8 Anxiety 149
9 PTSD, Depression, and Sleep 163
10 Psychosis and Schizophrenia 181
11 Addiction 195
12 Conclusion 207

References 221
Index 301

# Preface

The writing of this book has been a labor of love for all three authors, having been drawn together by our keen interest in the science of cannabidiol (CBD). The first paper that Raphael Mechoulam published in the cannabinoid field (among nearly five hundred lifetime publications) was the identification of the structure of the CBD molecule (Mechoulam and Shvo 1963), which allowed its synthesis and an evaluation of its mechanism of action in several biological assays. For the past sixty years, several groups have collaborated with Mechoulam, conducting some of the earliest studies on the potential medicinal benefits of this compound. Among those collaborations was one with Linda Parker's laboratory in Canada, investigating the effects of cannabinoids on nausea, vomiting, anxiety, pain, and addiction in pre-clinical rodent models (see Mechoulam and Parker 2013). Erin Rock joined in this collaboration, first as an undergraduate at Wilfrid Laurier University (Waterloo, Ontario) and then as a graduate master's and PhD student at the University of Guelph, identifying the mechanism of action of the antinausea and antivomiting effects of CBD for her PhD research (Rock et al. 2012). Rock continued as a postdoctoral fellow/research associate with Parker, continuing to unlock the mysteries of CBD and several other cannabinoids using these models.

People have used the cannabis plant for millennia for its medicinal and mind-altering effects. This complex plant contains over 100 plant cannabinoids, including the most well known, $\Delta^9$-tetrahydrocannabinol (THC) and cannabidiol (CBD). Of the over 100 cannabinoid compounds in the cannabis plant, THC has been identified as essentially the only psychotropic compound, based on research by Raphael Mechoulam's group in Israel and several others in the 1960s and 1970s. CBD, however, is not mind-altering.

Awareness of the potentially beneficial effects of CBD has grown at an astonishing rate in the mind of the general public, with Google Internet searches doubling in frequency every year for the past five years, and it is continuing to accelerate (Leas et al. 2019). Indeed, CBD has become a trendy ingredient in mass market products that make broad and at times unsubstantiated claims of its ability to treat a myriad of symptoms from skin disorders to chronic pain, as well as cosmetic use—and in many cases, without human clinical trial evidence. Many pet owners are also administering CBD for management of conditions such as pain and anxiety without relevant scientific evidence for these indications.

This current "CBD craze" often generalizes to human health on the basis of findings in cells or in preclinical rodent research. However, human clinical trial research has severely lagged behind the basic cellular and preclinical animal research on the beneficial effects of CBD, with the exception of the use of CBD in rare forms of childhood epilepsy. The lack of clinical trial data is surprising, given that over sixty years ago, small-scale human trials for treatment of epilepsy, addiction, and anxiety showed that CBD may be a promising potential treatment option. However, the regulatory rules governing research with cannabis, a Schedule I drug, prohibited large-scale research with humans on the therapeutic potential of CBD. In recent years with countries such as Canada and several US states legalizing cannabis, one would expect that access to CBD has become much more feasible for large-scale human clinical trials. However, at the time of writing this book, this has not yet been the case. As consumers have increased access to a variety of cannabis products, US and Canadian scientists face the burden of strict regulatory scrutiny (Haney 2020) and have a limited variety of cannabis and CBD to evaluate in trials. Despite these barriers, a current survey of the National Institutes of Health website, www.clinicaltrials.gov, revealed 276 planned, ongoing, or completed human trials with CBD (the vast majority using oral formulations) for many of the indications that have shown promise in the preclinical research.

It must be emphasized that the standardized, chemically pure CBD available for preclinical research and human clinical trials is not necessarily the consumer CBD available for sale from vendors and the Internet. A 2017 survey (Bonn-Miller, Banks, and Sebree 2017) reported that of eighty-four online CBD and hemp oil products examined, only twenty-six were accurately labeled for CBD and THC content, with CBD often being overlabeled and THC underlabeled, consistent with warnings from the Food and Drug Administration. Buyer beware!

Many drugs used today are natural products or their derivatives. So far, CBD has been approved as a treatment by the US Food and Drug Administration only for some rare forms of childhood epilepsy and seizures associated with tuberous sclerosis complex in patients one year of age or older. In this book, we discuss various aspects of CBD's actions. In many disease states, mostly in animal models, but also some in human studies, positive results have been noted and published. In view of the encouraging animal as well as the limited human clinical data and the relatively low level of toxicity or major side effects, we expect that CBD or, more likely, CBD derivatives with an improved pharmacological profile may be developed as drugs to treat several additional medical conditions in the future.

# 1
# Introduction

*Cannabis sativa* may be one of the oldest cultivated plants, yet its use has been associated with controversy throughout its long history (Russo, Guy, and Robson 2007). This historical use of cannabis provides clues to the potential treatment for an array of medical syndromes that remain challenging for twenty-first-century medicine. Cannabis has been used for millennia in the making of textiles, for casual recreational purposes, in religious practices, and in medicine, but its use has been associated with legal and societal controversies. In ancient China, cannabis was prescribed for several diseases but was noted to lead to "seeing devils" when taken in excess (Mechoulam et al. 2014)

We now know that the intoxicating effects produced by cannabis use are caused by the compound $\Delta^9$-tetrahydrocannabinol (THC), first identified in the 1960s by the young chemists Yechiel Gaoni and Raphael Mechoulam at the Weizmann Institute in Israel at that time (Gaoni and Mechoulam 1964). This discovery allowed scientists around the world to investigate the psychotropic or mind-altering effects of cannabis in the laboratory. The identification of THC led to the surprising discovery of a previously unknown neurochemical system called the endocannabinoid system, with *endo*, meaning inside, and *cannabinoid*, meaning compound of the cannabis plant. This system, including the cannabis-like compounds that we produce in our body, is now known to be intimately involved in key regulatory functions of human health and disease states. Even if you do not use cannabis, this system is present and always active in your body, constantly working to keep your internal systems in balance. It acts as a switch to return your body to its optimal internal state.

Even before the discovery of THC, another compound found in cannabis, cannabidiol (CBD), had been identified in the 1940s by Roger Adams in the United States and Alexander Todd in the United Kingdom (Adams, Hunt, and Clark 1940; Jacobs and Todd 1940). But no further work on CBD was published until 1963 when Raphael Mechoulam identified its structure (Mechoulam and Shvo 1963). No one reported on the activity or effects of CBD until the early 1970s, except to determine that it had no intoxicating properties. Since then, a great deal of research has been focused on determining the effects of CBD in the body and how it might be having these effects. A review of the journal database program PubMed reveals a dramatically accelerating base of CBD research over the past forty years, with over thirteen hundred research articles on CBD published in 2020 alone. (See figure 1.1 for a visual display of the research articles about CBD published in this database.)

Very early research in the 1970s and 1980s revealed that CBD may alleviate seizures in epileptic patients. Indeed, as reviewed in chapter 4 on epilepsy, cannabis has been historically used for the treatment of epilepsy. Over forty years ago, clinical trials demonstrated that CBD could be administered orally (at high doses) to alleviate seizures in epilepsy patients. This treatment received approval by the Food and Drug Administration (FDA) in 2018 for the treatment of a rare form of epilepsy in children, Dravet syndrome, which has a very high mortality rate and is unaffected by traditional treatments. Beginning in the first year of life, children affected by Dravet syndrome develop a severe brain disorder that results in cognitive,

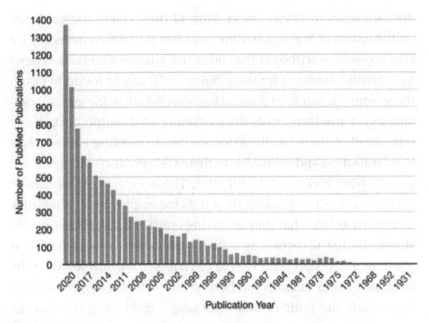

**Figure 1.1**
Articles published on CBD in the PubMed database over the years.

behavioral, and motor impairment. In 2016, research on the treatment of this devastating condition was accelerated by widespread press coverage of anecdotal successes of CBD-rich cannabis in treating Dravet syndrome by Sanjay Gupta, chief medical correspondent for the television channel CNN. These personal success stories inspired physicians to carry out the high-quality randomized clinical trials (RCTs) described in chapter 4 to properly demonstrate the efficacy of CBD treatment for this disorder. However, studies that investigate CBD for other medical conditions are limited in number and often lack scientific rigor, controls, and adequate sample size to be able to draw meaningful scientific conclusions (Britch, Babalonis, and Walsh 2021).

At the same time, we are currently in an era in which CBD is claimed to effectively treat almost any disorder, including anxiety, psychosis, sleep, pain and inflammation, stroke, neurological diseases, acne, and even cancer. Most of these claims are not based on the results of high-quality human clinical trial data necessary for FDA approval described at the end of this chapter. Our book reviews the scientific evidence for these claims. CBD can be found in everything from "health food drinks," lotions, and chewable gummy candies to pet products, with the marketing of these products being way ahead of the scientific evidence. Most troubling is the lack of standardization and validation of these CBD products. Many do not contain what is claimed on the label. This is especially problematic for those subject to random drug tests for employment when they incorrectly believe that they are taking only CBD. CBD is also poorly absorbed into the circulatory system when taken orally or rubbed as a lotion on the skin (as reviewed in chapter 2). Understanding the effect of CBD in humans has been greatly affected by government regulations and policies that can make it difficult to conduct the clinical trials that are needed to collect high-quality evidence for CBD's effects.

The gold standard seal of approval of a drug for the treatment of a medical condition is in its rigorous approval by the FDA. Pure oral CBD (Epidiolex) for rare forms of childhood epilepsy discussed in chapter 4 received FDA approval in 2018. Many drugs begin the FDA approval process, but considerably fewer actually gain approval. Figure 1.2 presents a schematic of the process of FDA approval. Each ascending level represents a more rigorous form of evidence supporting the efficacy of the treatment for a particular indication in humans.

At the base of the pyramid are basic laboratory studies—in vitro studies. In vitro (meaning in the glass) studies are performed with microorganisms, cells, or biological organisms outside the human or animal body. The compound is applied to these organisms or cells, and measures such as molecular changes, toxicity, or binding affinity are taken. With in vitro evidence that a treatment like CBD effectively modifies cellular signals coincident with a specific disease, testing then moves up the pyramid to preclinical in vivo studies with

Introduction    5

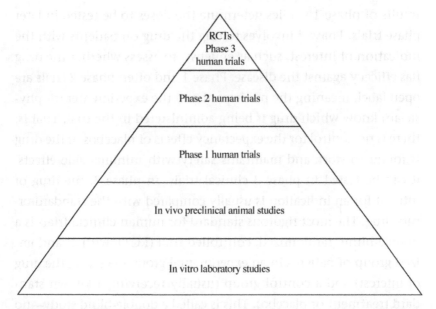

**Figure 1.2**
Schematic of FDA approval process

animal models of the disease. In vivo (within the living organism) work is done with a whole, living organism; common laboratory species are mice and rats. With appropriate controls in place, these studies can be conducted to infer cause and effect in a specific animal model, such as an animal model of epilepsy. Such studies are scrupulously vetted by institutional animal care committees, in compliance with national and international standards of animal care for laboratory animals to ensure minimal distress for the animal in light of the benefits of the experimental question for human health. These preclinical tests are required before a new treatment can be considered for testing in humans.

Human clinical trials then occur in three phases. Phase 1 trials, the first in humans, involve testing multiple doses of the drug on healthy volunteers for safety and often pharmacokinetic properties (discussed in chapter 2). The doses that are evaluated are usually only a fraction of the dose tested in preclinical trials to ensure safety. The results of phase 1 studies determine the doses to be tested in later phase trials. Phase 2 involves testing the drug on patients with the indication of interest, such as epilepsy, to assess whether the drug has efficacy against the disease. Phase 1 and often phase 2 trials are open label, meaning the participant and the experimenter (or physician) know which drug is being administered in the trial. That is, there is no control for the expectancy effects of placebos. If the drug is found to work and maintains safety with minimal side effects, it can be tested in phase 3 clinical trials. In phase 3, the drug of interest for an indication is usually compared with the standard-of-care drug. The most rigorous standard for human clinical trials is a double-blind, randomized, controlled trial (RCT), with a randomized group of patients in an experimental group (receiving the drug of interest) and a control group (usually receiving a known standard treatment or placebo). This is called a double-blind study—no one knows who is assigned to which group—and ensures against expectancy effects. Participants, caregivers, outcome assessors, and analysts are all candidates for being blinded. To assess whether a drug is beneficial, a comparison between the experimental group and the control group is examined for the outcome(s) of interest. In the approval process, the FDA evaluates a drug's clinical benefit and its risk information, with continued monitoring after approval.

This is the process that the pure oral form of CBD, Epidiolex, went through in order to be prescribed by physicians to treat patients with Dravet syndrome and Lennox-Gastaut syndrome, two rare forms of childhood epilepsy. The FDA has since also approved Epidiolex for the treatment of seizures associated with tuberous sclerosis complex in patients one year of age or older. This means that the FDA has concluded that it is safe and effective for this intended use. Following its approval in 2018, Epidiolex was designated as schedule 5 (the least restrictive category for medications with low abuse potential) and has subsequently been descheduled (meaning that it is no longer a controlled substance under the federal Controlled Substances Act).

## Legal Status of Cannabis and Cannabis-Derived Products in the United States

Over the past two decades, the legal status of cannabis in the United States has substantially changed. Prior to 1996, cannabis was not legalized for use for any purpose in the United States. California became the first state to legalize cannabis use for medical purposes. As of publication of this book, thirty-three US states have legalized adult use of cannabis for medical purposes. Eleven states have legalized its adult recreational use. In the November 2020 US election, four additional states voted to enact legalization of recreational use.

Federally, cannabis currently remains classified as a Schedule I substance in the United States, which means that it has no current acceptable medical use and has a high abuse potential. Access to a Schedule I substance is highly controlled for researchers, and the cannabis that is authorized is often not representative of the cannabis products that are commercially available. However, the legal status is likely to change from the time of publication of this book, with the US Congress currently discussing federal legalization of cannabis.

The legal status of CBD has been intricately linked with that of cannabis. The Agriculture Improvement Act of 2018 (also known as the Farm Bill) removed hemp (defined as plant material having less than 0.3 percent THC) from the list of controlled substances in the Controlled Substance Act, meaning that if CBD is derived from hemp, it does not fall within the jurisdiction of the US Drug Enforcement Administration (DEA). This means that any CBD derived from cannabis plants that contain more than 0.3 percent THC remains Schedule I. The 2018 Farm Bill preserves the FDA's regulation of products derived from cannabis, including CBD. Under these regulations, because CBD is listed as an active ingredient in an approved pharmaceutical drug, it cannot be added to food products and dietary supplements. In the eyes of the FDA, this would be akin to your favorite barista adding acetaminophen to your morning espresso, which is not permitted according to federal law. Consumers are rightly confused because CBD is commercially available to them in many products.

Clinical studies using CBD must still be approved by the FDA. To protect human participants, the FDA regulates how the cannabis product is cultivated, manufactured, and tested. These requirements have made it difficult to conduct the properly controlled RCTs that are needed. Therefore, it is ironic that as consumers have greater access to CBD, US scientists face greater regulatory scrutiny, including federal and state Schedule I licenses, and extensive regulations on storing and dispensing Schedule I drugs (Haney 2020). FDA regulations are appropriately cautious about what scientists can test in patients, and none of the CBD products that are available online or in dispensaries have gone through the safety and manufacturing procedures necessary for FDA approval. How can researchers proceed under such strict regulations? It has been suggested (Haney 2020) that one step to addressing some of the barriers to cannabinoid researchers is to give scientists a Schedule I exemption in order to increase the number of RCTs to provide empirical evidence of efficacy of cannabinoid compounds. As well, policy-oriented and regulatory research to guide rules about advertising, labeling, and evaluating the effect of different formulations of cannabis products is needed.

## The Cannabis Act in Canada

With the Cannabis Act, Canada legalized cannabis for recreational use in 2018, leaving researchers hopeful that Canada could become a key player in cannabis RCTs. With the act, patients authorized by their health care provider to use cannabis for medical purposes were authorized to buy product from a federally licensed seller or register to produce their own medical cannabis (or designate someone to produce it for them).

Unfortunately, despite the urgent need for researchers to conduct RCTs on cannabis products for medical purposes, the boom in Canadian cannabis research that was expected did not happen. This is mostly because of how clinical trials have been regulated by Health Canada. The barriers facing cannabis research include research funding prioritizing intervention research and the onerous processes involved in obtaining research licenses from Health Canada. At this time, most medical cannabis products that are available to purchase cannot be used in research trials because clinical trial products have more stringent manufacturing practice standards. This means it is not possible to conduct RCTs on the medical cannabis products that are currently being used by Canadians. Canada's opportunity to be a leader in medical cannabis research has so far been lost.

The barriers hindering researchers from conducting RCTs have affected cannabis research in general, and certainly research on its constituent, CBD. In each chapter of this book, for each indication discussed, we review the levels of evidence available for CBD from in vitro to RCTs.

# 2
# Chemical and Pharmacological Aspects

Cannabis contains a mixture of many closely related compounds. These compounds were difficult to identify until recently with the development of modern methods, such as chromatography and spectrometry. Chromatography is needed to isolate the chemical substances in a mixture, and spectrometry is needed to then determine that substance's chemical makeup and structure. Over the past few decades, 554 compounds in cannabis have been identified, including 113 phytocannabinoids and 120 terpenes (aromatic oils that give cannabis strains their distinctive scents; Ahmed et al. 2008; ElSohly and Gul 2014).

Over the past ten years, the concentration of THC has been increasing in the illicit market, but recently more high-CBD products have also been produced. Of interest is a recent report (ElSohly et al. 2021) on the concentration of seven major cannabinoids, including $\Delta^9$-THC and CBD, in illicit herbal cannabis products seized by the Drug Enforcement Agency in the last ten years in the United States. The number of samples seized decreased dramatically over the past five to six years because of legalization of marijuana for either medical or recreational purposes in many states. Of the confiscated samples analyzed, the mean $\Delta^9$-THC concentration increased from 10 percent in 2009 to 14 percent in 2019, with the mean THC:CBD ratio increasing from 25 in 2009 to 105 in 2017. Perhaps because of the evolving interest in the potential benefits of CBD, the THC:CBD ratio in 2019 decreased to 25, suggesting a trend in the production of more high-CBD products.

The therapeutic potential of these cannabis constituents has been mostly limited to the investigation of THC and CBD, leaving many unanswered questions regarding the biological activity of most of the other cannabinoids found in cannabis. Indeed, there has been considerable discussion of "entourage effects" (Mechoulam et al. 2014) among these cannabinoids; that is, the effect of cannabis is more than just the sum of its parts. Synergistic or at times antagonistic effects occur because there are different actions of the various cannabinoids at specific receptors in the body. It is like an orchestra that can blend and mix its components in an almost unlimited fashion. The emphasis of this book, however, is to identify the current state of knowledge about the effects of CBD as one of these components of cannabis. In this chapter, we describe the chemistry and pharmacology of CBD.

## Cannabidiol: Chemical Aspects

Cannabidiol (CBD) was isolated in 1940 from marijuana in the laboratory of Roger Adams in the United States (Adams, Hunt, and Clark 1940) and from Indian hemp resin in the laboratory of Alexander Todd in the United Kingdom (Jacobs and Todd 1940). They suggested a tentative partial structure based on chemical degradation and correlation with the known cannabinoid constituent cannabinol. In 1963, the correct full structure was reported by Mechoulam and Shvo (1963) and in 1964 by Santavy (1964), mostly based on nuclear magnetic resonance (NMR) data. The absolute stereochemistry was determined in 1967 (Mechoulam and Gaoni 1967a).

CBD is a crystalline compound with a melting point of 66°C to 67°C. Its NMR spectroscopy and infrared spectra are presented in a review (Mechoulam and Gaoni 1967b). The crystalline structure of CBD was determined by Jones et al. (1977). The aromatic ring and the terpene ring were found to be almost perpendicular to each other. This is in contrast to the structure of $\Delta^9$-tetrahydrocannabinol (THC), the main psychoactive cannabis constituent, in which the two rings are almost in the same plane. However, the different conformations in the crystalline state may be irrelevant to the difference in activity between CBD and THC, as the two rings in CBD can freely rotate in solution and in the gaseous state.

## Synthesis of CBD

Of the several CBD syntheses that have been reported (Jung et al. 2019), the most efficient one seems to be the acid condensation of p-mentha-2,8-dien-1-ol with olivetol, as originally proposed by Petrzilka and colleagues (1967) and later improved (Baek, Srebnik, and Mechoulam 1985; figure 2.1). The yield reported (41 percent) in this one-step reaction makes CBD readily available. In addition to CBD, this reaction also leads to the abnormal CBD (abn-CBD), a relatively unstudied cannabinoid.

## Chemical Reactivity of CBD

**Acidic Conditions** Under strong acidic conditions (heating with p-toluene sulphonic acid; p-TSA) CBD is converted first into $\Delta^9$-THC, which is then isomerized to $\Delta^8$-THC. Under different acidic conditions (with $BF_3$), CBD is ring closed into $\Delta^9$-THC, as well as into $\Delta^8$-iso-THC (figure 2.2) (Gaoni and Mechoulam 1966a, 1966b). While $\Delta^9$-THC has

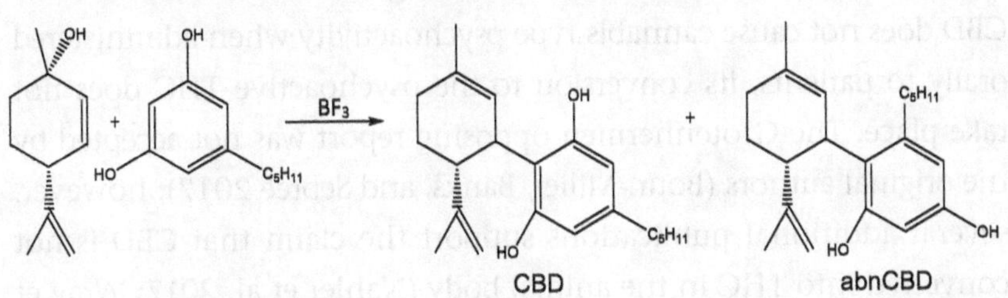

**Figure 2.1**
The synthesis of CBD

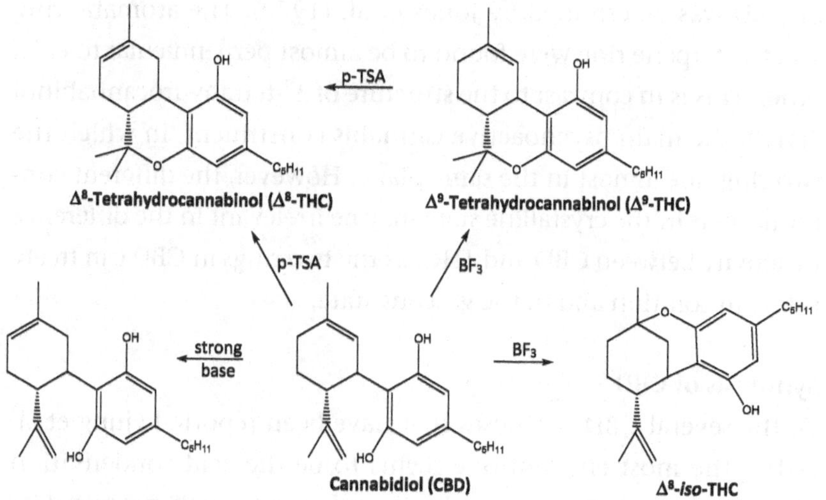

**Figure 2.2**
Reactions of CBD under acidic and basic conditions

been the subject of hundreds of publications, $\Delta^8$-iso-THC, which is also psychoactive, has been almost totally neglected.

Recently it was claimed that CBD can convert into $\Delta^9$-THC upon prolonged exposure to simulated in vitro acidic gastric conditions (Merrick et al. 2016). This report was strongly opposed by Grotenhermen and colleagues (2017), who claimed that the conversion of CBD into THC does not take place in vivo in the stomach and brought evidence that the in vitro conditions may not be representative of the in vivo gut environment and that this reaction does not seem to be relevant to the actual in vivo conditions that exist in our gut. A main point that Grotenhermen and colleagues stressed is that because CBD does not cause cannabis-type psychoactivity when administered orally to patients, its conversion to the psychoactive THC does not take place. The Grotenhermen opposing report was not accepted by the original authors (Bonn-Miller, Banks, and Sebree 2017); however, several additional publications support the claim that CBD is not converted into THC in the animal body (Nahler et al. 2017; Wray et al. 2017; Crippa et al. 2020).

**Basic Conditions** Under specific basic conditions (potassium in tert-pentyl alcohol), CBD has been isomerized to its $\Delta^6$ analog (Srebnik et al. 1984), which has received very little investigation (figure 2.2). In order for the isomerization to take place, both phenolic groups have to be free. Apparently the CBD phenolate is involved in this reaction. Contrary to the natural CBD ($\Delta^1$ isomer), the synthetic $\Delta^6$ isomer of CBD showed THC-like activity in rhesus monkeys (Mechoulam and Hanus 2002). Again, as with the $\Delta^8$-iso-THC, mentioned above, while the plant CBD ($\Delta^1$ isomer) has been the topic of many hundreds of publications, its $\Delta^6$ isomer has not received any attention.

**Oxidation** CBD in base in the presence of oxygen is oxidized to the p-quinone (figure 2.3), a potent topoisomerase II inhibitor (Kogan

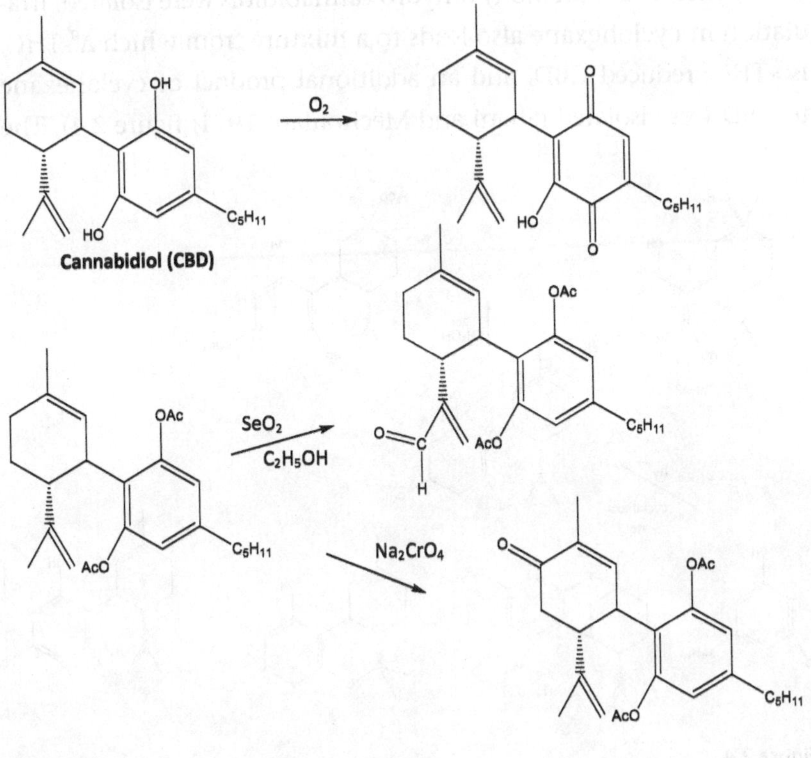

**Figure 2.3**
Oxidations of CBD and CBD diacetate

et al. 2004; Kogan, Schlesinger, Priel, et al. 2007). A comparative in vivo study in mice has shown that the CBD quinone is less toxic and more effective in reducing tumor growth than the widely used anticancer drug doxorubicin, which is also a topoisomerase II inhibitor (Kogan, Schlesinger, Peters, et al. 2007). Oxidation of CBD diacetate with selenium dioxide leads to the aldehyde on the C-10 position, while oxidation with sodium chromate takes place on the C-6 position (Lander et al. 1976; figure 2.3). In several studies, CBD has been found to possess potent antioxidant activity, which may be the basis of its assumed role in neurodegenerative diseases (Hacke et al. 2019).

## Photochemical Reactions of CBD

Irradiation of CBD in methanol with a 450 W lamp gives a mixture from which two 1-methoxy dihydro cannabidiols were isolated. Irradiation in cyclohexane also leads to a mixture from which Δ⁹-THC, iso-THC, reduced CBD, and an additional product of cyclohexane to CBD were isolated (Shani and Mechoulam 1971; figure 2.4). The

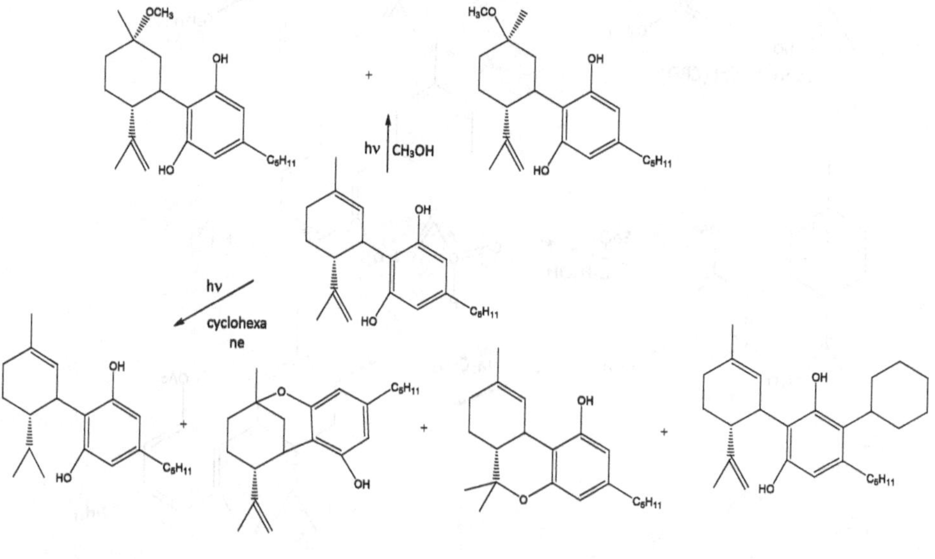

**Figure 2.4**
Photochemical reactions of CBD

instability of CBD on irradiation (possibly only in solution) should be taken into account in laboratory studies as well as in commercial products.

## (+) Cannabidiol

The unnatural (+) CBD enantiomer and some of its derivatives have been reported (Leite et al. 1982; Hanus et al. 2005). Surprisingly, contrary to the compounds in the (–) series, which bind very weakly to the $CB_1$ receptor, (+) CBD and most of the derivatives in the (+) series bind to the $CB_1$ receptor in the nanomole range, as well as to the vanilloid receptor type 1 (VR1) (Hanus et al. 2005; Bisogno et al. 2001). Some of these compounds also bind weakly to the $CB_2$ receptor. A derivative of (+) CBD was reported to be more active than the (–) enantiomer in an ovulation blockage assay in rats (Cordova et al. 1980).

## Cannabidiolic Acid

CBD is actually not a natural product. The cannabis plant produces the unstable cannabidiolic acid (CBDA), which is nonenzymatically decarboxylated to CBD (figure 2.5) often by heating or drying of the plant material. Preclinical investigations in rats indicate that CBDA is 100 to 1,000 times more potent than CBD in reducing nausea and vomiting (Bolognini et al. 2013; Rock and Parker 2013), stress-induced anxiety (Bolognini et al. 2013; Rock et al. 2017), and inflammatory pain (Rock, Limebeer, and Parker 2018). At much higher doses, it also seems to parallel CBD in its antiepileptic activity (Anderson et al. 2019). However, CBDA is unstable and may not be optimal as a medicine. Recently it was shown that the methyl ester of CBDA (HU-580) is stable, and several reports on its biological activities have been published. It can cause suppression of nausea and stress-induced anxiety in rats, which is serotonin 1A (5-$HT_{1A}$) receptor-mediated (Pertwee et al. 2018). In animal models, it also lowers depression-like behavior (Hen-Shoval et al. 2018), as well as neuropathic pain (in male animals only!) (Zhu et al. 2020).

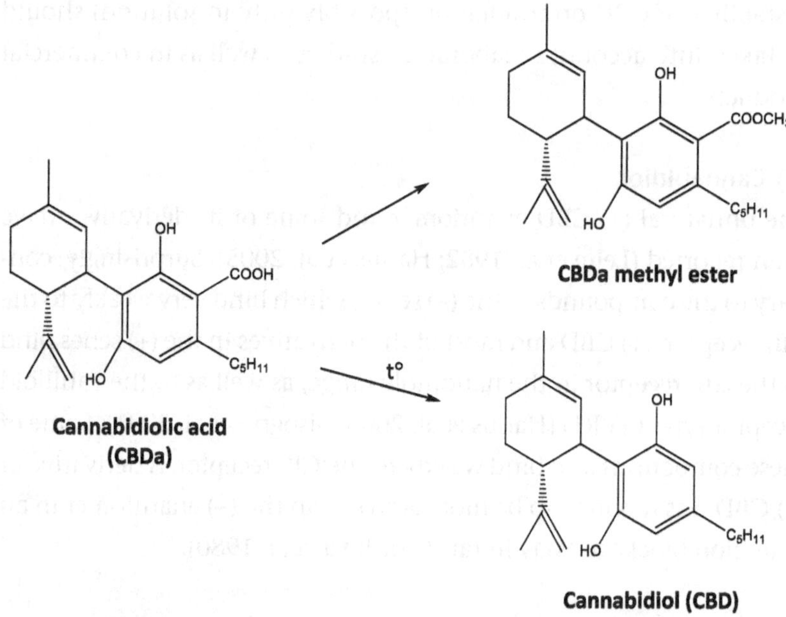

**Figure 2.5**
Reactions of cannabidiolic acid

Murillo-Rodríguez and colleagues (2020) have recently reported that the methyl ester of CBDA prolonged wakefulness and decreased slow-wave sleep duration. It enhanced extracellular levels of dopamine and serotonin in the nucleus accumbens, while adenosine and acetylcholine were increased in the basal forebrain.

## Derivatives of Cannabidiol

Many drugs used today are semisynthetic (made from the synthesis of a natural product) derivatives of natural products, developed in order to enhance the activity of the natural product or to make it more suitable for human use. Derivatives of penicillin and cortisone are typical examples. Derivatives of CBD have also been reported. Thus, 4'-F-CBD, was prepared by direct fluoridation of CBD (Breuer et al. 2016; figure 2.6). It was found to be considerably more potent than CBD in behavioral assays in mice predictive of anxiolytic,

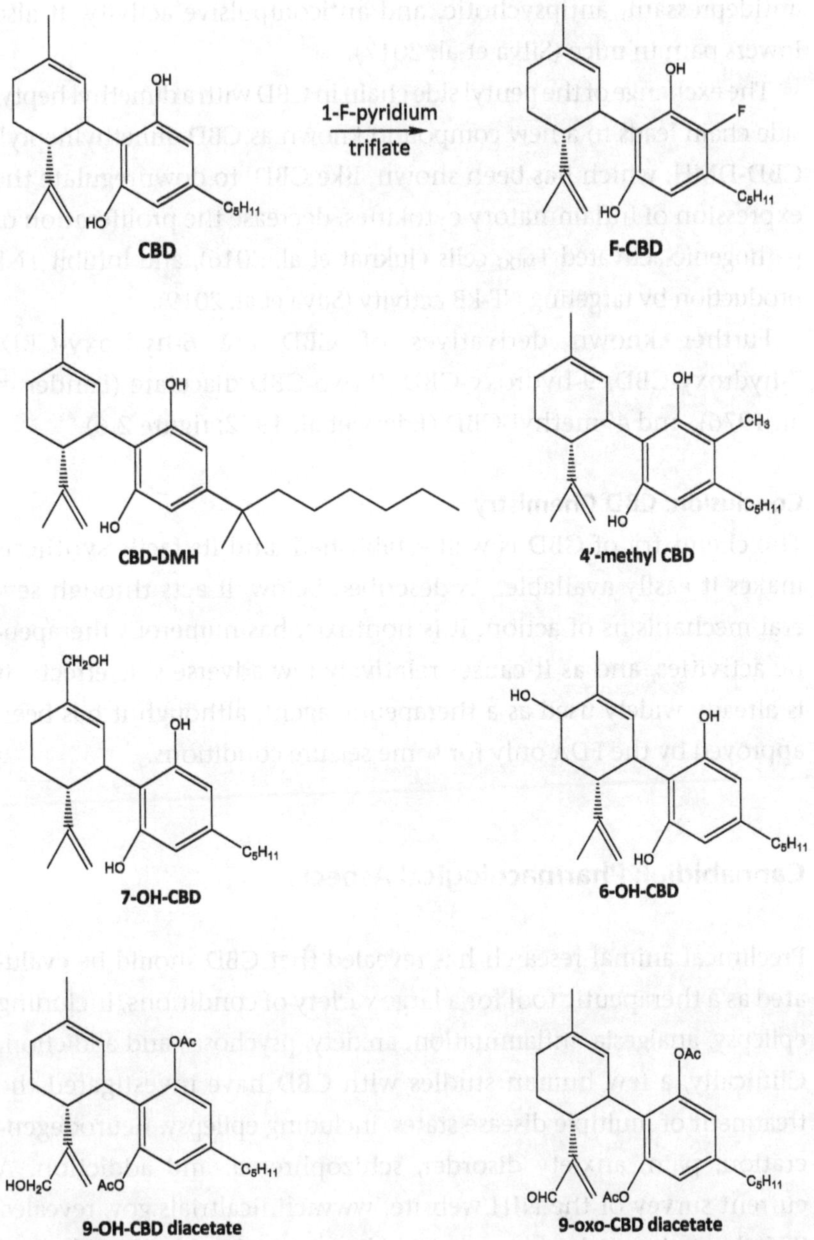

**Figure 2.6**
CBD derivatives

antidepressant, antipsychotic, and anticompulsive activity. It also lowers pain in mice (Silva et al. 2017).

The exchange of the pentyl side chain in CBD with a dimethyl heptyl side chain leads to a new compound known as CBD-dimethylheptyl, CBD-DMH, which has been shown, like CBD, to downregulate the expression of inflammatory cytokines, decrease the proliferation of pathogenic-activated $T_{MOG}$ cells (Juknat et al. 2016), and inhibit TNF production by targeting NF-kB activity (Silva et al. 2019).

Further known derivatives of CBD are 6-hydroxy-CBD, 7-hydroxy-CBD, 9-hydroxy-CBD, 9-oxo-CBD diacetate (Lander et al. 1976), and 4'-methyl-CBD (Edery et al. 1972; figure 2.6).

**Conclusion: CBD Chemistry**

The chemistry of CBD is well established, and its facile synthesis makes it easily available. As described below, it acts through several mechanisms of action. It is nontoxic, has numerous therapeutic activities, and as it causes relatively few adverse side effects, it is already widely used as a therapeutic agent, although it has been approved by the FDA only for some seizure conditions.

**Cannabidiol: Pharmacological Aspects**

Preclinical animal research has revealed that CBD should be evaluated as a therapeutic tool for a large variety of conditions, including epilepsy, analgesia, inflammation, anxiety, psychosis, and addiction. Clinically, a few human studies with CBD have investigated the treatment of multiple disease states, including epilepsy, neurodegeneration, pain, anxiety disorder, schizophrenia, and addiction. A current survey of the NIH website, www.clinicaltrials.gov, revealed 276 planned, ongoing, or completed human trials with CBD (the vast majority using oral formulations) for many of the indications that have shown promise in the preclinical research. Over-the-counter

CBD is also marketed as an unregulated food supplement/medication. It has been reported (Bonn-Miller et al. 2017) that 69 percent of these unregulated products marketed as "CBD-only" were inaccurately labeled (43 percent underlabeled for CBD concentration, 26 percent overlabeled for CBD concentration, and 21 percent contained THC [up to 6.4 mg/L]). Despite the prevalence of CBD use, guidance on dose recommendations has not advanced, mostly because of the lack of pharmacokinetic and bioavailability data for CBD in humans. Only a few published studies report oral bioavailability of CBD in humans. As well, there are few dose-determination studies, limiting our understanding of the desired plasma concentrations to achieve minimum effective doses. There is a lack of information about how different formulations and routes of administration affect absorption into the bloodstream. We review the published data available below.

## CBD Bioavailability

The term *bioavailability* refers to the fraction of an administered dose of unchanged drug that reaches the systemic circulation. When a drug is administered directly into the bloodstream intravenously, its bioavailability is 100 percent. When a drug is administered orally, inhaled, or delivered transdermally (by application to the skin), its bioavailability decreases because of incomplete absorption into the blood or by first-pass metabolism by enzymes in the liver. It is clear that the bioavailability of CBD varies with the route of administration.

Bioavailability is assessed by a number of metrics reflecting pharmacokinetic (PK) properties of drugs. $C_{max}$ means the maximum (peak) concentration in the blood recorded, and $T_{max}$ means the time it takes to reach that peak concentration. The area under the plasma-drug concentration-time curve (AUC) is expressed as ng/L for CBD. The AUC reflects that actual body exposure to the drug after administration of a given dose and is dependent on the absorption and elimination of the drug from the body. The elimination half-life of a drug is defined as the time it takes for the concentration of the drug in the plasma to be reduced by 50 percent; that is, after one-half-life, the concentration of the drug in the body will be half of the starting dose. These metrics are used to reveal the bioavailability of drugs. CBD is generally administered orally as either a capsule or dissolved in an oil solution (e.g., olive or sesame oil) in clinical trials or research studies. It can also be administered through sublingual, intranasal, and transdermal routes; however, the relative plasma absorption of these routes is not well understood.

*Smoking/vaporizing.* Smoking CBD-rich hemp or a CBD-biased cannabis strain provides a rapid and efficient delivery from the lungs to the body. The bioavailability of aerosolized CBD has been reported to yield rapid peak plasma concentrations in five to ten minutes and higher bioavailability (approximately 31 percent) than oral administration (Grotenhermen 2003; Ohlsson et al. 1986). Because vaporizers (commonly called e-cigarettes) typically operate at temperatures that do not combust the cannabis product being inhaled, they expose cannabis smokers to fewer toxicants (e.g., carbon monoxide) compared to smoked methods (Spindle, Bonn-Miller, and Vandrey 2019). Vaporized cannabis users report fewer respiratory symptoms compared to those who predominantly smoke cannabis, but the long-term health effects of regular users of smoked versus vaporized cannabis remain unclear (Newmeyer et al. 2016). In 2019 there was an outbreak of emergency department visits in the United States related to vaping THC products, but the incidence has declined due to the removal of vitamin E acetate from the products. Laboratory data show that vitamin E acetate, an additive in some THC containing vaping products, was strongly linked to these outbreaks (Blount et al. 2020). However, it is not known if long-term use of vaporizers may be harmful.

*Topical formulations.* Some of the most popular emerging cannabis products are topical, with formulations including balms, creams, salves, gels, and patches. At this time, there are no controlled clinical studies or representative surveys with which to assess the pharmacokinetic profile, behavioral effects, and use characteristic of these products (Spindle, Bonn-Miller, and Vandrey 2019). Few data are available about the bioavailability of topical formulations. In dogs, CBD (from Applied Basic Science Corporation) was administered by transdermal cream, oral CBD-infused oil, or oral CBD in microencapsulated beads (Bartner et al. 2018) at doses of 750 mg and 1,500 mg and CBD levels in plasma were measured over the first twelve hours. Over the first twelve hours, bioavailability was eight times higher by oral infused oil than by transdermal cream at a dose of 75 mg and three times higher at a dose of 150 mg. Since CBD is highly lipophilic, it accumulates in the outer layer of the skin (strata corneum) and does not penetrate through the skin layers. New delivery technologies are being developed to enhance permeation approximately three-fold, as tested in guinea pigs (Paudel et al. 2010). New research (described in detail in chapter 11) indicates that transdermal CBD (combined with a permeation enhancer) at twenty-four-hour intervals for seven days reduced relapse to alcohol or cocaine self-administration in rats, a model of addiction, as well as anxiety in an elevated plus maze (Gonzales-Cuevas et al. 2018).

*Oral administration.* The most common method for administration of CBD is orally or by submucosal spray. The vast majority of clinical trials posted on www.clinicaltrials.gov involve oral administration of CBD. CBD is a highly lipophilic drug, with its solubility in the aqueous environment of the gut being in the range of only several milligrams per liter (Samara, Bialer, and Mechoulam 1988). This low water solubility leads to incomplete absorption. As well, when administered orally, CBD undergoes first-pass metabolism in the liver (Martin, Harvey, and Paton 1977), while a large proportion (33 percent) is excreted unchanged in the feces. The earliest work with humans estimated the bioavailability of oral delivery of 20 mg CBD to be only 6 percent due to significant first-pass metabolism in the liver (Agurell et al. 1981). Subsequently, 900 mg of CBD was administered orally to a monkey, and extremely low levels of plasma CBD were detected over the entire experiment (Jones et al. 1981). In dogs, oral CBD (180 mg by gelatin capsule) was detected in the blood of only three out of six dogs with bioavailability values of 13 percent, 13 percent, and 19 percent, respectively (Samara, Bialer, and Mechoulam 1988).

GW Pharmaceuticals' plant-derived highly purified oral CBD (Epidiolex, 100 mg/ml) was FDA approved in the United States in June 2018 for seizures associated with Lennox-Gastaut syndrome (LGS) or Dravet syndrome (DS) in patients two years of age and older. Four randomized, placebo-controlled trials (Devinsky et al. 2017; Devinsky, Patel, Cross, Villanueva, Wirrell, Privitera, Greenwood, Roberts, Checketts, VanLandingham, Zuberi, et al. 2018; Devinsky, Patel, Thiele, et al. 2018; Thiele et al. 2018) demonstrated the efficacy and safety of CBD in these severe and treatment refractory epilepsies. Recently, a phase 2 human clinical trial with both a single ascending dose and multiple-dose pharmacokinetic trial (Taylor et al. 2018) of this oral formulation of CBD (Epidiolex) was reported in healthy adult volunteers. The single ascending dose arm of the study was double-blind, randomized, and placebo controlled. Four groups of healthy participants received single ascending oral doses of 1,500, 3,000, 4,500, or 6,000 mg CBD or matching placebo. Over the course of CBD treatment, the $C_{max}$ of CBD and metabolites increased with a trend to less than dose proportionality; that is, as the dose increased, the $C_{max}$ did not increase representatively. CBD and metabolites appeared rapidly in the plasma, with a $T_{max}$ of four to five hours independent of the dose. The effective half-life of CBD was in the range of ten to seventeen hours, independent of the dose. Given the high lipophilicity of CBD, solubility-limited absorption of higher doses is a likely explanation for the lack of dose-dependent elevation of $C_{max}$. Lim, Sharan, and Woo (2020) reviewed several studies using single dose (5,000–6,000 mg) oral CBD plasma concentration-time profiles in healthy participants with the conclusion that CBD absorption saturated around 4,000 mg, where the amount absorbed into the body (dose multiplied by bioavailability) approaches its plateau. When absorption saturation occurs, systemic exposure does not proportionately increase with increasing doses.

In the multiple dose arm of the phase 1 double-blind, randomized, and placebo-controlled trial (Taylor et al. 2018), two groups received multiple oral doses of 750 or 1,500 mg CBD or matching placebo twice daily, under fasted conditions for seven days. Steady-state CBD plasma concentrations were reached at approximately two days. At steady state, there was an almost doubling in exposure for a twofold increase in dose (750–1500 mg) with $C_{max}$ increasing by about 1.6-fold and AUC increasing approximately 1.9-fold. Therefore, unlike single dosing, multiple dosing increased bioavailability in almost a dose-proportional manner, at least at these two lower doses. At steady state, $T_{max}$ of CBD and its metabolites occurred at approximately three hours, independent of the dose. Single doses up to 6,000 mg and multiple doses up to 1,500 mg twice daily were well tolerated, with only mild and moderate aversive effects reported (diarrhea, nausea, headache, and somnolence) during the trial. There were no discontinuations due to aversive effects.

Finally, the phase 1 trial (Taylor et al. 2018) also included a food effect arm, which was randomized to a period (fed [high fat breakfast], then fasted or fasted, then fed). Participants were dosed thirty minutes after starting breakfast and then fasted for four hours. In the fed state, bioavailability of CBD was increased in all groups (4.85-fold increase in $C_{max}$; 4.2-fold increase in AUC) compared to the fasted state. The bioavailability of CBD is clearly increased by food. However, there was no effect of food on $T_{max}$ or on half-life of CBD or its metabolites. This effect was recently confirmed (Crockett et al. 2020) following a single oral 750 mg dose of CBD (Epidiolex in healthy adults following a high-fat/calorie meal ($n=15$), a low fat/calorie meal ($n=14$), whole milk ($n=15$), or alcohol ($n=15$) relative to a fasted state ($n=29$). Blood samples were collected until ninety-six hours postdose and analyzed by liquid chromatography and tandem mass spectrometry. The bioavailability of CBD administered with a high fat/calorie meal increased by 5.2-fold versus fasted, low fat/calorie meal by 3.8-fold versus fasted, whole milk by 3.1-fold versus fasted and alcohol by 3.1-fold versus fasted. The meal state did not affect $T_{max}$ of CBD. As well, Birnbaum et al. (2019) evaluated the pharmacokinetics of purified oral CBD capsule with and without food in adults with refractory epilepsy. A single dose of 99 percent pure CBD capsules was taken in fasting (no breakfast) and fed (840–860 calorie high-fat breakfast) conditions. Blood sampling for CBD in plasma concentrations was performed under each condition between 0 and 72 hours postdose and measured by liquid chromatography mass spectrometry assay. Bioavailability of CBD was fourteen times higher in the fed state than in the fasted state. No adverse effects were reported.

Clearly, in a fasted state, oral CBD has poor bioavailability, but in a fed state, bioavailability is significantly improved. One of the current focuses of pharmaceutical research is the development of new vehicles (solutions that CBD is mixed in) for delivery of CBD to increase bioavailability. Such vehicles include capsules in which CBD is suspended in solutions of long-chain fatty acids high in oleic acid (Patrician et al. 2019) or in nano-lipospheres of less than 60 nm in diameter (Atsmon et al. 2018), both showing higher bioavailability relative to generic CBD.

Once in the bloodstream, CBD is rapidly distributed into tissue and, like THC, may preferentially accumulate in adipose tissues due to its high lipophilicity.

**CBD Metabolism**

CBD is extensively metabolized in the liver (Martin, Harvey, and Paton 1977). The primary route is hydroxylation to 7-OH-CBD, which is then metabolized further, resulting in a number of metabolites that are excreted in feces and urine. In vitro studies with human liver microsomes (Jiang et al. 2013, 2011) demonstrated that seven human cytochrome P450 (CYP) enzymes are capable of metabolizing

CBD, with the two main forms being CYP3A4 and CYP2C19, known to also be involved in the metabolism of various clinically important drugs. This suggests that CBD may interact with other drugs by this mechanism; in fact, there is a report that the metabolism of $\Delta^9$-THC is inhibited by CBD (Jones and Pertwee 1972). A recent report (Taylor et al. 2019) suggests that exposure to CBD metabolites was increased in subjects with moderate and severe hepatic impairment, suggesting that dose modification is necessary in such patients—they should start at a lower dose with slower titration. However, CBD was well tolerated in this population as well. In the recent clinical trials for epilepsy, CBD is given as an adjunct treatment along with the patient's prescribed antiepilepsy medication(s) (typically three medications), which may interfere with drug-drug metabolism. It has been shown that concomitant administration of CBD with clobazam (a common antiepilepsy drug) produced increased plasma concentrations of clobazam's major active metabolite, N-desmethylclobazam, likely the result of CBD inhibition of CYP2C19, which may result in the adverse event of sedation, a common side effect of clobazam administration (Geffrey et al. 2015; Gaston et al. 2017). However, a recently reported clinical trial (Gaston et al. 2019) showed that such drug-drug interactions cannot account for the effects of CBD on reducing seizure frequency and severity in treatment-resistant epilepsy. Interestingly, data suggest that acute treatment with CBD (120 mg/kg, i.p.) inactivated at least one cytochrome P450 isozyme in mice livers, but after repetitive CBD (120 mg/kg, i.p. for four days) treatment, an isozyme was induced that was resistant to further doses of CBD (Bornheim and Correia 1989). This isozyme appears to be similar to the isozyme induced by phenobarbital (a common anticonvulsant medication that controls epileptic seizures) in mice. A comprehensive review (Balachandran, Elsohly, and Hill 2021) describes the interactions between CBD and other medications, illicit substances, and alcohol.

## Can CBD Be Converted in Vivo to THC?

In the laboratory using certain methods, CBD can be converted to THC (Mechoulam and Hanus 2002); however, the likelihood that CBD converts to THC in the body of CBD consumers is very low. There were in vitro reports that CBD could be converted to THC spontaneously in the presence of an acid that could occur in the human gut (Merrick et al. 2016); however, subsequent research has indicated that this finding is limited to specific experimental conditions and likely does not occur when oral CBD is consumed by humans. A recent study supported by GW pharmaceuticals, the source of Epidiolex, examined the gastric and plasma concentrations of cannabinoids in mini-pigs (with a GI composite similar to humans) after repeated CBD administration (15 mg/kg/day for five days). No THC or THC metabolites were detected in plasma or gastric fluid after CBD administration (Wray et al. 2017). Generally there is no experimental evidence that this transformation occurs in humans after oral CBD administration. In one human study, 600 mg of CBD was administered to healthy participants. No THC or trace concentrations of THC metabolites (11-OH-THC, THC-COOH) were detected (Martin-Santos et al. 2012). Even chronic administration of CBD does not result in detectable THC concentrations in plasma; for example, in a six-week clinical study with Huntington's disease patients who were administered CBD 10 mg/kg/day (about 700 mg/day), CBD average plasma concentration range was 5.9 to 11.2 ng/ml with no THC detected (Consroe et al. 1991). In addition, a recent study (Kintz 2021) tested the blood of eight volunteers after vaping an e-cigarette containing 100 mg/mL of CBD, and no THC or THC-COOH was detected (at fifteen and forty-five minutes after the last use). This is of importance, as the blood testing was achieved under acidic conditions, suggesting no evidence of in vivo conversion of CBD to THC when CBD was administered by vaporization. Most recently, McCartney et al. (2021) reported that orally administered CBD did not produce false positive tests for THC on standard point of collection oral fluid testing devices. In a randomized, double-blind, crossover design, healthy participants ($n = 17$) completed four treatment sessions involving administration of placebo, 15 mg, 300 mg, or 1500 mg pure CBD in a high-fat dietary supplement. Oral fluid samples were tested at baseline, 20, 145, and 185 minutes post-treatment. THC was not detected in any samples. Recent reviews have summarized a number of studies involving high doses of CBD consistently failing to demonstrate THC-like effects (psychomotor impairment, increased HR, tachycardia, dry mouth; Nahler et al. 2017; Grotenherman, Russo, and Zuardi 2017). Overall, there is no evidence that oral CBD administration in humans results in clinically relevant THC-like subjective or physiological effects or appreciable plasma concentrations of THC or its metabolites.

## Impact of CBD on Drug Testing for THC

The impact of CBD exposure on urine drug testing for THC has not been well studied. Spindle et al. (2020) characterized the urinary pharmacokinetic profile of 100 mg oral and vaporized CBD, vaporized CBD-dominant cannabis (100 mg CBD; 3.7 mg $\Delta^9$-THC), and placebo in six healthy adults using a within-subjects crossover design. Urinary peak concentrations of CBD were higher after oral (mean $C_{max}$:776 ng/ml) versus vaporized CBD (mean $C_{max}$: 261 ng/ml). CBD concentrations peaked five hours after oral CBD ingestion and within one hour after inhalation of vaporized CBD.

Due to the increased use of CBD and CBD-dominant cannabis products, there is a need to understand the impact of CBD on urine drug-testing programs commonly used in the workplace, criminal justice, drug treatment, and other settings. Urine is the primary biological choice for drug testing, and the most commonly targeted analyte to evaluate cannabis exposure is 11-nor-9-carboxy-$\Delta^9$-tetrahydrocannabinol ($\Delta^9$-THCOOH), a metabolite of $\Delta^9$-THC. Although drug testing is not aimed at detecting CBD in existing drug-testing procedures, contaminated CBD products could theoretically produce a positive result for $\Delta^9$-THCOOH on a urine drug test. Indeed, given the nature of the unregulated cannabis industry, products may falsely advertise as containing only CBD but may also contain $\Delta^9$-THC in concentrations ranging from trace levels to those capable of producing intoxication (and detection in drug testing). As well, hemp-derived CBD products can contain up to 0.3 percent $\Delta^9$-THC, and even the FDA-approved CBD medication, Epidiolex, may contain trace levels (less than 0.1 percent) of $\Delta^9$-THC, and possibly increase the risk of testing positive for THC. Acute oral ingestion or inhalation of a 100 mg dose of CBD did not result in positive urine drug test when using screening and confirmatory cutoffs in the mandatory guidelines for federal workplace drug testing in the United States (Spindle et al. 2020). In contrast, inhalation of cannabis containing 100 mg CBD and 3.7 mg $\Delta^9$-THC resulted in positive test results for two of the six participants. Urinary concentrations of CBD were higher and peaked later when CBD was orally ingested compared to when it was inhaled. There was no evidence in this study that CBD converts to $\Delta^8$-THC or $\Delta^9$-THC in the human gut in a fed state. However, it is not known if this would also not occur in a fasted state when the gut is more acidic.

**CBD Pharmacodynamics**

CBD has been reported to have several molecular sites of action for its various effects. A summary of CBD pharmacodynamics is presented in table 2.1. Some have argued that the myriad potential therapeutic effects of CBD may be related to these varied mechanisms of action.

CBD does not appear to act directly at $CB_1$ or $CB_2$ receptors—there is very little measurable response in binding assays; that is, only at high concentrations (greater than 10 µM) has CBD been found to bind with $CB_1$ receptors or $CB_2$ receptors (Pertwee 2008). However, using human $CB_2$ receptors in HEK293A cells, CBD has more recently been reported to act as a partial agonist (in the nM range; Tham et al. 2019). It appears to act indirectly at both cannabinoid receptors. It

**Table 2.1**

CBD pharmacodynamics

| Receptors | Ion Channels | Enzymes | Other |
|---|---|---|---|
| Low $CB_1/CB_2$ affinity $CB_1$- or $CB_2$-negative allosteric modulator | Activation of TRPV1, TRPV2, TRPV3, and TRPM8 | FAAH inhibitor Inhibits CYP1A1, 2B6, 2C19, 3A4, and 3A5 | Binds to FABPs (competitive inhibition) Adenosine uptake inhibitor |
| $5HT_{1A}$ agonist | Calcium channel inhibition | Cyclooxygenase inhibitor/activator | |
| Peroxisome proliferator-activated receptor gamma (PPAR γ) activator | | | |
| GPR55 antagonist | | | |
| GPR18 agonist Inverse agonist for GPR3, GPR6, and GPR12 | | | |
| Adenosine receptor agonist | | | |

Abbreviations: $5HT_{1A}$, serotonin1A; $CB_1$, cannabinoid receptor type 1; $CB_2$, cannabinoid receptor type 2; CBD, cannabidiol; CYP, cytochrome P450; FAAH, fatty acid amide hydrolase; FABP, fatty acid binding protein; GPR, g- protein coupled receptor; TRPM, melastatin-related transient receptor potential cation channels; TRPV1, transient receptor potential cation channel subfamily V member.

is a negative allosteric modulator of the $CB_1$ receptor, thereby acting as a noncompetitive antagonist of the actions of THC and other $CB_1$ agonists (Laprairie et al. 2015). CBD also has been reported to act as a negative allosteric modulator at the $CB_2$ receptor (Martinez-Pinilla et al. 2017). In rodent studies, CBD may also enhance the action of anandamide, an endogenous ligand of the $CB_1$ and $CB_2$ receptors by blocking its degradation by the enzyme, fatty acid amide hydrolase (FAAH) (Bisogno et al. 2001); however, this mechanism may not occur in humans (Criscuolo et al. 2020). CBD dampens nitric oxide (NO) production in animal models of inflammation and the expression of inflammatory cytokines and transcription factors and reactive oxygen species (ROS) production. However, in cancer cells, CBD is capable of generating ROS, thereby inducing cytotoxicity or apoptosis and autophagy (Ligresti et al. 2006).

CBD has also been shown to modulate several non-endocannabinoid neurotransmitter signaling systems that may be responsible for its therapeutic effects. Considerable evidence suggests that CBD enhances serotonergic activity at the 5-$HT_{1A}$ receptor (Rock et al. 2012; Russo, Burnett, and Hall, 2005) which may mediate its antinausea, antiemetic, and anxiolytic effects. It has also been shown to act on several other neurotransmitter systems, including inhibition of the uptake of adenosine, agonism of several channels belonging to the transient receptor potential (TRP) family (TRPV1, TRPV2, TRPA1), enhancement of the activity of glycine receptor subtypes, activation of peroxidase proliferator–activated receptor γ (PPARγ), and blockade of the orphan G-protein coupled receptor GPR55 (Pertwee 2008). It has been suggested that such a multitarget action by CBD is key for its diverse therapeutic potential for several indications. The potential mechanisms of action of CBD for each of the therapeutic end points discussed in this book are reviewed in the relevant chapter.

## CBD Safety and Toxicology

Several reviews have evaluated the safety and toxicity of CBD (Iffland and Grotenhermen 2017; Bergamaschi, Queiroz, Zuardi et al. 2011). Generally, in vitro and preclinical animal studies have found CBD to be relatively low in toxicity; however, less is understood about the toxicity of long-term use in humans, especially of the high therapeutic doses used for many indications such as psychosis (see chapter 10). There are also no human data on the effects of CBD in pregnancy and in breastfeeding mothers on fetal and infant development. CBD modifies growth of tumoral cell lines but has no effect on most nontumor cells (Massi et al. 2006). Although the evidence is severely limited, CBD does not appear to affect the development of embryonic cells (Paria, Das, and Dey 1995). In preclinical animal research (Rosenkrantz, Fleischman, and Grant 1981), CBD has generally no effects on a wide range of physiological and biochemical parameters or on animal behavior on its own unless extremely high doses are given (e.g., in excess of 150 mg/kg i.v. as an acute dose or exceeds 30 mg/kg orally daily for ninety days in monkeys). The effect on the immune system is unclear, with evidence of immune suppression at higher concentrations but immune stimulation at lower concentrations.

There is a potential for CBD to be associated with drug interactions through inhibition of some cytochrome P450 enzymes (Sholler, Schoene, and Spindle 2020; Balachandran, Elsohly, and Hill 2021). The impact of CBD alone, and in combination with other medications, on liver function needs continued monitoring. Specifically, CBD may inhibit metabolic activity of several cytochrome (CYP) P450 drug-metabolizing liver enzymes that are instrumental in the metabolism of many common prescription and over-the-counter drugs. More specifically, administration of CBD in combination with drugs metabolized by certain CYP enzymes (for example, CYP2C9, CYP2D6) may lead to delayed metabolism of these drugs. For example, a study of children with refractory epilepsy found that administration of Epidiolex (5–25 mg/kg/day) in combination with clobazam (antiseizure medication) elicited about a 500 percent increase in plasma concentrations of the active metabolite of clobazam (norclobazam; Geffrey et al. 2015). In a recent review (Balachandran, Elsohly, and Hill 2021), CBD has been reported to interact with antiepileptic drugs, antidepressants, opioid analgesics, and THC, but it surprisingly also interacted with several other common medications, including acetaminophen and alcohol. Continued research is needed in this area.

In contrast to THC, CBD has no effect on heart rate or blood pressure under normal conditions, but in animal models of stress, it reduces heart rate and blood pressure (Sultan et al. 2017). Several clinical trials with humans have assessed safety and toxicology of CBD as a therapy for treatment-resistant epilepsy. A thorough review of these studies (Iffland and Grotenhermen 2017) revealed that the most commonly reported side effects were fatigue, diarrhea, and changes of weight or appetite. Since CBD is most commonly used as an adjunct therapy, more clinical research on the impact of CBD on hepatic liver enzymes, interactions with other drugs, and effects on embryonic development is needed.

A more recent meta-analysis of randomized clinical trials with CBD (Chesney et al., 2021) suggests that the likelihood of voluntary withdrawal from the study was greater compared with placebo. The likelihood of withdrawal was also a function of CBD dosage, with low-dose CBD (20–400 mg/day) not differing from placebo. Serious adverse effects were only reported at high doses of CBD (1400–3000 mg/day), particularly in studies of epilepsy in children with the participants taking other epilepsy medications that can interact with CBD. The doses of CBD provided by health and food supplements are typically much lower (5–20 mg/day), so incidence of adverse events is likely to be much lower; however, this has yet to be demonstrated. This issue requires more research as some over-the-counter products are poorly labeled or do not contain purified CBD, and may possibly contain other cannabinoids and contaminants.

**CBD Dependence and Abuse Potential**
Controlled human studies regarding the potential physiological dependence/tolerance effects of CBD in humans have not been reported. In preclinical animal studies, however, CBD does not appear to produce dependence. Male mice were injected with CBD (0.1, 1, or 3 mg/kg) or THC (1, 3, 10 mg/kg) daily for fourteen days to determine tolerance to the neuroprotective effects of these compounds against cerebral ischemia. Tolerance developed to the neuroprotective effects of THC but not CBD over the course of treatment (Hayakawa et al. 2007). As well, tolerance does not develop to the antinausea effects of CBD over seven days of treatment (Rock, Sullivan, Collins, et al. 2020). Several preclinical animal studies have shown that CBD lacks abuse potential, using a variety of animal models, including a reduction in threshold frequency for intracranial

self-stimulation (ICSS), dopamine release in the mesolimbic system, conditioned place preference, and drug discrimination. A low dose (5 mg/kg) of CBD did not modify the threshold frequency required for ICSS; however, high doses of CBD (10 and 20 mg/kg) elevated the threshold frequency for ICSS, opposite to that of drugs of abuse (such as amphetamine, cocaine, and opiates; Katsidoni, Anagnostou, and Panagis 2013). Unlike most drugs of abuse, CBD did not modify the dopamine release in cells of the reward system, consisting of the mesolimbic ventral tegmental area-nucleus accumbens pathway (French, Dillon, and Wu 1997). CBD (10 mg/kg) alone produces neither a conditioned place preference nor a conditioned place aversion. However, rats treated with increasing doses of CBD and THC (1, 3, and 10 mg/kg) showed a trend toward a conditioned place preference not seen with THC alone (Klein et al. 2011), which may represent a pharmacokinetic interaction leading to higher concentration of THC rather than a receptor action change. CBD does not exhibit THC-like discriminative stimulus effects in rats trained to discriminate between THC and vehicle (Vann et al. 2008). CBD clearly does not produce rewarding or aversive effects on its own in preclinical animal models.

Human studies of the abuse potential of CBD are limited. However, a randomized double-blind, placebo-controlled trial study indicated that CBD, unlike THC, did not have abuse potential following a single oral dose of 600 mg of CBD in healthy volunteers (Martin-Santos et al. 2012). Unlike CBD, THC (10 mg oral) was associated with subjective intoxication, euphoria, sedation, hallucinogenic activity, increased psychotic symptoms, and anxiety. As well, although THC increased heart rate, CBD did not. In another randomized double-blind, within-subject laboratory study, CBD (0, 200, 400, 800 mg, oral) combined with inactive (0.01 percent THC) and active (5.3–5.8 percent THC) smoked cannabis, was assessed for abuse potential in healthy cannabis smokers. The participants completed eight outpatient sessions with CBD administered ninety minutes prior to cannabis self-administration. Among the groups administered 0 mg/kg CBD (placebo conditions), active cannabis was self-administered by significantly more participants and produced time-dependent increases in subjective ratings and heart rate relative to inactive cannabis. CBD produced no significant psychoactive, cardiovascular, or other effects among the inactive cannabis conditions. Active cannabis self-administration, subjective effects, and cannabis ratings did not vary as a function of CBD dose relative to placebo conditions. These findings suggest that oral CBD does not reduce the reinforcing, physiological, or positive subjective effects of smoked cannabis.

## Conclusion: Pharmacological Aspects of CBD

CBD has diverse molecular targets including indirect activity at the cannabinoid receptors, agonism of TRPV and 5-HT$_{1A}$ receptors, and additional receptor targets are being investigated. Some argue that the many potential health benefits are related to the widespread action of CBD. The consensus is that pharmaceutical-grade CBD has a favorable safety profile with limited side effects, but this may not generalize across all populations or across all CBD formulations, as unregulated retail products carry inaccurate labels. As well, there are limited reports that daily chronic administration of CBD may inhibit activity of liver CYP 450 drug-metabolizing enzymes. More research on interactions of CBD with other drugs is crucial.

The most common form of CBD treatment is oral administration, which is plagued by low bioavailability because of first-pass metabolism and acidic stomach conditions, requiring considerably high doses to achieve therapeutic levels. Human trials have revealed, however, that repeated treatments with lower doses produce a steady state of CBD availability when it is administered with food. The bioavailability of topically applied CBD has been little studied, whereas bioavailability is higher when CBD is smoked or vaporized. CBD is metabolized in the liver by CYP 450 enzymes, with the most common forms being CYP3A4 and CYP2C19, known to be involved in the metabolism of various clinically important drugs. This suggests that CBD may produce drug interactions through this mechanism. CBD may also interfere with some effects of THC by this mechanism. CBD, unlike THC, does not appear to have abuse potential and has relatively fewer adverse side effects; it is not intoxicating. CBD acts on several receptors, which we review in the following chapters regarding its mechanism of action for each indication.

# 3
# CBD/THC Interactions

Unlike THC, CBD does not appear to have intoxicating effects, but it may interact with THC when coadministered to either counteract or potentiate the effects of THC (Boggs, Nguyen, et al. 2018, Freeman et al. 2019). A number of preclinical animal and clinical human studies have shown that CBD may ameliorate some of the adverse effects of THC (Osborne, Solowij, and Weston-Green 2017). In human studies of cannabis users, cannabis with higher levels of CBD (assessed by user hair analysis, analysis of plant material, or estimation of proportional exposure) has been associated with better cognitive performance, especially memory (Morgan et al. 2018; Morgan, Schafer, et al. 2010), and with fewer psychotic symptoms (Morgan and Curran 2008, Schubart et al. 2011). An open label trial of prolonged CBD administration in humans (200 mg/day) for ten weeks showed improved psychological symptoms and cognition and increased hippocampal subfield volume in cannabis users (Beale et al. 2018; Solowij et al. 2018). Controlled administration studies have shown that pretreatment with oral CBD reduced the cognitively impairing and paranoia-inducing effects of intravenously administered THC (Englund et al. 2013), and simultaneous infusion of THC and CBD blocked THC related anxiety and subjective alterations (Zuardi et al. 1982).

The cultivation of high THC potency strains of cannabis involved the selective breeding out of CBD in the plant matter, such that until recently, very low levels (or none) of CBD were present in typical street cannabis (ElSohly et al. 2016). Indeed, it has been speculated that the absence of CBD in cannabis of high THC potency may contribute to the psychosis-like outcomes (Di Forti et al. 2009) and lack of protection from harm to the brain (Yucel et al. 2016). Some investigators have recommended breeding CBD back into the cannabis plant material and cannabis products as a harm-reduction strategy and to maximize therapeutic benefits (Englund et al. 2017). More recent samples of confiscated marijuana by the DEA indicate that the ratio of THC:CBD decreased from 104 in 2017 to 25 in 2019 (ElSohly et al. 2021), suggesting that more marijuana is being produced with higher CBD concentrations.

## Pharmacodynamic Effects

It is conceivable that these CBD/THC interactions occur within the endocannabinoid system. THC is a partial agonist at $CB_1$ and $CB_2$ receptors, while CBD is a low-affinity $CB_1$ and $CB_2$ receptor ligand and negative allosteric modulator of $CB_1$ (Laprairie et al. 2015) and $CB_2$ (Martinez-Pinilla et al. 2017), reducing the binding of agonists (such as THC or anandamide), while augmenting endocannabinoid tone through inhibition of FAAH (Bisogno et al. 2001), ultimately elevating levels of fatty acids such as anandamide. However, the effect of CBD on FAAH inhibition has recently been suggested to be exclusively evident in rodent, not human, cells (Criscuolo et al. 2020). The diametrically opposing actions of CBD and THC on $CB_1$ and $CB_2$ receptors provide a potential pharmacological mechanism for in vivo interactions between these compounds. However, CBD also acts at receptors other than cannabinoid receptors (eg. 5-HT$_{1A}$, TRPV1, adenosine), further complicating how it may functionally interact with THC.

**Pharmacokinetic Effects**

It is possible that the change in pharmacokinetic metabolism of THC due to the presence or absence of CBD may explain some of the interactive effects. The direction (potentiation or attenuation) of the CBD/THC interactions in preclinical models may depend on the relative times of CBD and THC administration. When CBD is administered thirty minutes (or even up to 24 hours) prior to THC, it has been shown to potentiate the effects of THC in rats, whereas coadministration can result in blockade of THC effects (Zuardi, Hallak, and Crippa 2012). The potential of these pretreatment intervals to potentiate THC effects may be related to an increase in the effective brain exposure to THC because CBD has been shown to inhibit the liver enzyme P450, which metabolizes THC (Jones and Pertwee 1972). Coadministration of equal doses of CBD:THC nearly doubled brain THC levels when collected thirty minutes after i.p. injection (Klein et al. 2011). Therefore, preclinical animal studies seem to suggest that CBD administration prior to THC may result in higher THC levels (due to CBD inhibiting its metabolism by the liver), leading to a more pronounced THC effect.

To date, eight studies have assessed the pharmacokinetics of CBD/THC interactions in humans, with most reporting that CBD does not significantly alter the pharmacokinetic profile of THC. However, three of these studies did suggest that CBD may have a small effect on THC metabolism (Freeman et al. 2019). A study of fourteen participants indicated that a combination of vaporized THC (13.75 mg) and vaporized CBD (13.75 mg) produced higher peak plasma concentrations of THC when compared to vaporized THC (13.75

mg) alone (Arkell, Kevin et al. 2019). Coadministration of oral CBD (5.4 mg) altered the metabolism of oral THC (10 mg) by partial inhibition of the cytochrome P450 enzymes, which hydroxylate THC to its metabolite 11-hydroxy-THC (Nadulski et al. 2005); however, there was wide variability among the twenty-four participants. CBD combined with THC produced higher levels of THC metabolites among women compared to men, which was not produced by THC alone (Roser et al. 2009). A study with thirty-six participants found no significant difference in plasma concentration of THC following pretreatment with oral CBD (600 mg) or placebo (Englund et al. 2013). A small crossover study with six participants found that pretreatment with intravenous CBD (5 mg) did not influence THC blood levels following intravenous THC (1.25 mg) when compared to placebo (Bhattacharyya et al. 2010). Oral administration of 1500 mg CBD did not alter the pharmacokinetics of i.v. THC (Hunt et al. 1981). Finally, oral THC (5 and 15 mg) produced similar maximum plasma levels of mucosal nabiximols (CBD/THC mixtures) at CBD: THC ratios of 5:5.2 and 15:16.2 mg (Karschner, Darwin, Goodwin et al. 2011). The limited data currently available suggest that CBD may have a very small effect, if at all, on THC metabolism in human participants. However, the widespread intraindividual variability, different routes of administration, and often different timing of CBD-THC administration, as well as different blood collection timing following drug exposure, makes cross-study comparison problematical.

## Preclinical Evidence of THC/CBD Interactions

The preclinical animal literature contains considerable evidence of CBD/THC interactions, with some suggesting that CBD reduces THC's adverse effects, such as memory deficits (Wright, Vandewater, and Taffe 2013) and others suggesting that CBD potentiates the beneficial effects of THC, such as prevention of nausea (Rock, Sullivan, Pravato et al. 2020). It has been suggested that this is partly dose related, with low doses of CBD potentiating THC effects and high doses of CBD reducing THC effects (Zuardi, Hallak, and Crippa 2012; Solowij et al. 2018). Indeed, it has been suggested that CBD:THC ratios need to be 8:1 to see antagonistic effects and only 2:1 to see potentiation effects in rodents (Zuardi, Hallak, and Crippa 2012). However, in monkeys, CBD reduced THC-induced behavioral impairments whether it was administered simultaneously or thirty minutes before THC (Wright, Vandewater, and Taffe 2013) and 1:1 to 3:1 ratios relative to THC were effective in attenuating the cognitive effects of THC in monkeys (Jacobs et al. 2016; Wright, Vandewater, and Taffe 2013). Yet a recent finding suggests that daily THC plus CBD does not interfere with the cognitive deficits produced by THC alone in adolescent nonhuman primates (Withey et al. 2021), although CBD did interfere with vomiting produced by acute high-dose THC (see chapter 6). Species differences may account for these varying effects.

Evidence also suggests that THC increases (French, Dillon, and Wu 1997), but CBD decreases the activity of mesocorticolimbic dopamine activity (Renard et al. 2016, 2017; Hudson et al. 2019). This opposing effect appears to be mediated by the activation of molecular pathways in the mesolimbic pathway by THC that may play a role in psychotic and addictive effects produced by THC that are counteracted by CBD.

A high dose of CBD (20 mg/kg) has been shown to reverse THC (1 mg/kg)-induced deficits in social interaction tasks in rodents (Malone, Jongejan, and Taylor 2009). As well, the antinociceptive effects of THC in mice or rats were enhanced by a very high dose (30 mg/kg, delivered i.v.) of CBD (Varvel et al. 2006), which was also accompanied by elevated THC blood levels, suggesting that CBD interfered with the metabolism of THC by interaction with P450 liver enzymes (Jones and Pertwee 1972). Combined CBD:THC in a 1:1 ratio both potentiated the locomotor suppressant effect of THC and interfered with the hypothermia induced by THC (Todd and Arnold 2016). Combined CBD (3 mg) and THC (3 mg) daily for three weeks during adolescence prevented the development of THC-induced cognitive and behavioral impairments in mice (Murphy et al. 2017). The inconsistent pattern of results across doses may a be a function of the biphasic effects of THC and CBD (Pertwee 2008). An inverted bell-shaped dose response curve for CBD has been reported in a number of acute administration studies with animals. For instance, low doses of CBD (2.5–10 mg/kg) but not higher doses reduced anxiety (Guimaraes et al. 1990) in animal models. As well, in humans administered the Simulated Public Speaking test, doses of CBD (100, 300, 900 mg) given orally showed an inverse bell-shaped curve with the medium dose (300 mg) showing strongest efficacy in reducing anxiety (Linares et al. 2019). Therefore, differential effects across studies may depend on the absolute dose, ratio of CBD:THC, route of administration, and timing of exposure to the combination and the species tested.

## THC/CBD Interaction in Humans

The evidence for CBD/THC interaction in humans is limited to cross-sectional population-based studies and a few clinical trials, with most research suggesting that CBD either attenuates or does not modify THC effects. However, caution should be used in interpreting many of these studies as they are based on correlational, not experimental, data.

### Pulse Rate and Blood Pressure

Ten studies have reported outcomes for the effects of combined CBD/THC on pulse rate, with eight studies reporting that THC alone and the combination of THC and CBD increased pulse rate when compared to baseline measurements or placebo, but there was no significant difference between THC alone and the combination of the two drugs (Freeman et al. 2019). In contrast, an early small parallel groups design ($n=5$/group) study reported results that oral THC (30 mg) alone significantly increased pulse rate. When that dose of THC was combined with oral CBD, a low dose of CBD (15 mg) potentiated the THC-induced increase in pulse rate, but higher doses of CBD (30 mg and 60 mg) prevented the effect of THC on pulse rate (Karniol et al. 1974). These differential effects of CBD/THC on pulse rate again show the biphasic dose effects of CBD.

Three studies reported outcomes of CBD/THC interactions on blood pressure, with two suggesting that CBD may alter the effect of THC on blood pressure (Freeman et al. 2019). One study found no significant changes in diastolic blood pressure, but three hours after drug administration, sublingual CBD (10 mg) prevented the decrease in systolic blood pressure produced by sublingual THC (10 mg) (Guy 2003). On the other hand, Freeman and colleagues (2018) reported increased systolic blood pressure produced by both sublingual THC (8 mg) alone and the combination of THC (8 mg) plus CBD (10 mg) when compared to placebo, but diastolic blood pressure increased only following THC alone, suggesting that CBD protected against the increase in diastolic blood pressure produced by THC. Finally, there were no blood pressure changes in a study of pretreatment with oral CBD (200, 400, 800 mg) or placebo ninety minutes before smoking THC (about 42 mg; Haney et al. 2016). Therefore, CBD may have a slight protective effect of blood pressure changes by THC, but the results are inconsistent.

**Subjective Intoxicating Effects**
Considerable variation exists among the studies that investigate the potential of CBD to modify the subjective intoxicating effects of THC. Most studies use visual analogue scales (VAS) where the participants self-rate levels of intoxication typically using a 10-point

scale anchored at "not stoned/high" and "extremely stoned/high." Among nine such studies, the most common finding is that CBD does not modify the subjective intoxicating effect of a given dose of THC (Freeman et al. 2019). However, three studies provide some evidence that CBD may reduce the acute subjective effects of THC. The first is a double-blind crossover study ($n = 15$) in which smoked THC (0.025 mg/kg) was associated with an increase in feelings of "high" and an increase in drug-related subjective effects; however, smoked CBD (0.150 mg/kg) significantly reduced these effects of THC (Dalton et al. 1976). Similarly, in a small double-blind crossover study, combined oral THC (0.5 mg/kg) and CBD (1 mg/kg) reduced the THC-induced subjective feelings on an Addiction Research Center Inventory for marijuana effects (Zuardi et al. 1982). Finally, a more recent randomized placebo-controlled trial of frequent and infrequent cannabis users examined the acute effects of these compounds alone and in combination when administered by vaporization (Solowij et al. 2019). Low doses of CBD when combined with THC enhanced, while high doses of CBD when combined with THC reduced, the intoxicating effect of THC. The enhancement of THC intoxication by low-dose CBD was particularly prominent in novice or less experienced cannabis users. These studies suggest that CBD may modify the intoxicating effects of THC in a subtle manner.

**Psychomotor Performance**

Psychomotor performance is critical in driving. Studies consistently find that CBD does not modify the psychomotor performance deficits produced by THC. Using a driving simulation task, both vaporized THC (13.75 mg) with and without CBD (13.75 mg) was found to increase lane weaving during a car-following task when compared with placebo (Arkell, Lintzeris et al. 2019). As well, oral CBD (0.320 mg/kg) did not modify the impairment produced by oral THC (0.215 mg/kg) alone in a reaction speed task, motor coordination task, and steadiness task (Bird et al. 1980). Finally, combined smoked THC (0.025 mg/kg) and CBD (0.150 mg/kg) and smoked THC (0.025 mg/kg) alone, when compared with placebo, reduced standing steadiness using a wobble board, hand-eye coordination on an attention motor performance task, and coordination and dexterity using a pegs test (Evans et al. 1976). Together these findings suggest that even cannabis strains with higher CBD content do not offset the impact that THC has on driving performance.

**Sleep**
Only one study has investigated the potential of CBD to modify the effect of THC on sleep in humans (Nicholson et al. 2004). The effect of oromucosal spray containing THC (15 mg) and CBD (15 mg) were compared to oromucosal spray containing only THC (15 mg). THC had sedative effects, but when combined with CBD, the duration of wakefulness and stage 3 nocturnal sleep were increased, suggesting that CBD may have activating properties. However, there were no differences in participants' subjective ratings about quality, duration, or timing of sleep onset. THC alone was also associated with increased subjective ratings of sleepiness thirty minutes after waking. CBD appears to mitigate against the sedative effects of THC, which coincides with reports that some users prefer to use CBD-rich strains of cannabis in the waking hours and THC-rich strains before bed (see chapter 9).

**Anxiety**
THC increases subjective effects of anxiety acutely in a biphasic dose-related manner. Lower doses decrease anxiety and higher doses increase anxiety. CBD, in contrast, may have antianxiety properties. The first of the few human clinical studies examining the interaction of THC and CBD on anxiety examined the effect of oral CBD alone and on THC-induced anxiety (Karniol et al. 1974). Oral CBD (15, 30, and 60 mg) did not modify anxiety on its own, but it reduced oral THC-induced anxiety when given simultaneously. Zuardi and

colleagues (1982) found that when oral CBD (1 mg/kg) was administered with oral THC (0.5 mg/kg), it attenuated the increases in anxiety symptoms associated with oral THC alone, as measured by the State-Trait Anxiety Inventory. More recently, Iain McGregor's laboratory in Australia (Arkell, Lintzeris, et al. 2019) reported that vaporized THC (13.75 mg) both with and without vaporized CBD (13.75 mg) produced increased ratings of anxiety on the State-Trait Anxiety Inventory after fifteen minutes, but only the THC-alone group maintained the increase in anxiety over sixty minutes. On the other hand, a study with approximately 1:1 CBD:THC (5 mg:5.4 mg or 15 mg:16.2 mg) failed to find a reduction of THC-induced anxiety by CBD (Karschner, Darwin, McMahon et al. 2011). See chapter 8 for further discussion.

## Cognition

Acute exposure to THC produces transient and dose-related cognitive impairments in humans (D'Souza et al. 2004), with the most robust effects on verbal learning, short-term memory, working memory, and attention. Some studies suggest that CBD may decrease the cognitive-impairment effects of THC, but the results are mixed. In cross-sectional studies, cannabis users completed verbal memory tasks when intoxicated and when not intoxicated (Morgan, Schafer, et al. 2010), and samples of cannabis smoked were assayed for the levels of THC and CBD. Specimens with higher levels of CBD were associated with better recall, suggesting a beneficial effect of CBD on cognition. A subsequent study examined verbal recall in heavy cannabis users or recreational users while not intoxicated. Hair samples were used to assay THC/CBD levels. Daily cannabis users with high hair THC concentrations performed worse on the verbal recall test. The presence of CBD was associated with better recognition recall (Morgan et al. 2012). These studies suggest that the presence of CBD in recreational cannabis may protect against the memory-impairing effects of THC.

These cross-sectional studies have limitations related to self-report regarding type of cannabis used and lack of randomization; that is, the individual characteristics determine the choice of cannabis to be used and may be related to the outcome. Experimental laboratory studies in which participants are randomly assigned to conditions can address these limitations. Healthy participants were pretreated with CBD (600 mg oral) or placebo prior to receiving intravenous THC (Englund et al. 2013). CBD protected against THC-induced verbal learning deficits. As well, a study examining the effects of THC (2.5 mg sublingual) and CBD (2.5 mg sublingual) in a clinical population of twenty-four people with a range of neurological symptoms, found that CBD reversed the THC-induced memory deficits. A large randomized double-blind, placebo-controlled crossover study of forty-eight cannabis users (twenty-four light and twenty-four heavy) examined the effects of oral CBD (16 mg), oral THC (8 mg), placebo, or the combination of THC and CBD on emotional facial recognition (Hindocha et al. 2015). CBD attenuated the THC-induced impaired facial recognition. Vaporized CBD (13.75 mg) did not prevent the memory-impairing effect of vaporized THC (13.75 mg) in an auditory memory task, the Paced Auditory Serial Addition Task (Arkell, Lintzeris, et al. 2019). A recent study (Cutler, LaFrance, and Stueber, 2021) also indicated that CBD added to high THC potency cannabis flower and cannabis concentrations did not offset deficits in memory and did not offset self-titration of doses with similar levels of intoxication. To the best of our understanding there have been no brain imaging studies to determine the effects of CBD/THC interactions during perceptual and cognitive tasks.

**Psychosis**
THC and high-potency THC cannabis can produce subjective effects that include paranoia, suspiciousness, perceptual alterations, and cognitive disorganization that can be measured using a number of standardized rating scales. In contrast to THC, CBD does not produce these subjective effects on its own and may have antipsychotic effects. The interaction of CBD/THC on psychosis-like symptoms is well studied in humans. In a web-based survey, information on the

amount and type of cannabis consumed was collected and psychiatric symptoms were evaluated using the Community Assessment of Psychic Experiences (CAPE). Self-reported use of cannabis with high CBD content was correlated with lower CAPE-positive symptoms, suggesting less psychiatric symptoms (Schubart et al. 2011). In population-based studies, greater psychosis proneness as assessed by the Oxford Liverpool Inventory of Life Experiences (Morgan and Curran 2008), and the Schizotypal Personality Questionnaire (Morgan et al. 2012) was observed in individuals using THC alone versus THC plus CBD. However, as these observations are based on a retrospective population analysis, it cannot be known if the individuals with higher psychosis proneness are simply more likely to use cannabis with lower CBD content. Human laboratory studies have investigated the effects of CBD on THC-induced acute psychotic-like effects. In three experimental studies, it was shown that THC produced positive psychotic symptoms in healthy volunteers randomly assigned to groups and pretreatment with CBD attenuated these effects (Bhattacharyya et al. 2010; Englund et al. 2013; Leweke et al. 2000). In contrast, a more recent investigation (Morgan et al. 2018) did not find any difference in acute psychotic symptoms between vaporized THC (8 mg) with or without vaporized CBD (16 mg); both groups displayed greater psychotic reactions than a placebo control. Although not entirely consistent, the bulk of experimental results suggest that CBD may reduce the psychosis-like effects of THC. In fact, CBD has recently been tested on its own as a therapy for schizophrenia (see chapter 10).

## Limitations of CBD/THC Interaction Studies

Although there is clear evidence for the potential of CBD to attenuate several acute and chronic adverse effects of THC, the data has limitations. Across studies, the route of administration (oral, sublingual, smoked, vaporized), the doses of THC and CBD, and the ratio of CBD:THC vary, making comparisons difficult. Therefore, the dose-response relationship influencing how CBD moderates the effect of THC cannot be determined. Several studies used oral administration, which has poor bioavailability of 13–19 percent (Mechoulam, Parker, and Gallily 2002), suggesting that these results should be interpreted with caution. As well the vast majority of the evidence comes from acute dosing, yet human use involves repeated exposure. In fact, Klein et al. (2011) found that chronic co-administration of CBD (3 mg/kg, i.p.) and THC (3 mg/kg, i.p.) in rats increased anxiety and reduced social interaction, compared with chronic THC alone. Additional studies of chronic dosing are warranted to support translational inferences about the interactive effects of CBD and THC. As well, there is very little work on inhalation of these compounds, yet that is the method of administration used by most cannabis users. Another limitation of the preclinical work is species differences. In primate models, there are clear parallels to human work showing that CBD restores THC's effects on memory (Wright, Vandewater, and Taffe 2013). In contrast, much of the rodent work suggests that CBD either fails to attenuate the effects of THC and often, instead, potentiates the effects (Boggs, Nguyen, et al. 2018).

**Nabiximols (Sativex)**
Nabiximols (Sativex), developed by G. W. Pharmaceuticals, is a specific cannabis extract botanical drug administered by oromucosal spray under the tongue. Each spray delivers a dose of 2.7 mg THC and 2.5 mg CBD. It was approved in 2010 in the United Kingdom to alleviate neuropathic pain and spasticity in multiple sclerosis (MS). It was subsequently approved in several European countries for this indication. In Canada, nabiximols has been approved by Health Canada for treatment of MS pain and spasticity, as well as cancer pain. To the best of our knowledge, nabiximols has not been approved by the FDA in the United States for any indication.

## Conclusion

Overall, the primarily correlational literature suggests that combining CBD with THC may reduce some of the potentially harmful effects of cannabis (memory impairment, anxiety, psychosis), but may not reduce its intoxicating effect or psychomotor deficits necessary for safe driving. More research is clearly necessary before we know if CBD-rich cannabis serves to reduce the harm produced by THC-rich cannabis. CBD may act to reduce some acute effects of THC through negative allosteric modulation of the $CB_1$ receptor (Laprairie et al. 2015), but there is also evidence that CBD may increase plasma concentrations of THC (Arkell, Lintzeris, et al. 2019). Clearly, future research aimed at establishing a clear dose-response function of both CBD, THC and their ratios using a common route of administration (e.g., vaporizing) used by cannabis users will be necessary before we can determine the appropriate dose and combined ratios of these cannabinoids in producing harmful or beneficial effects. The effect of repeated use is also crucial.

# 4
# Epilepsy

Epilepsy is a neurological disorder that manifests as recurrent, spontaneous seizures or convulsions with possible loss of consciousness due to disturbance of the excitatory-inhibitory equilibrium of neuronal activity in the brain. It affects approximately 1 percent of the world's population (Williams, Jones, and Whalley 2014). Two broad categories of seizures have been described: generalized and focal. Generalized seizures originate at a specific point within the brain but rapidly distribute across the brain to affect both hemispheres. Focal seizures are restricted to a specific region of the brain or a single hemisphere.

Conventional antiepileptic drugs may block sodium channels or calcium channels, or may enhance the inhibitory neurotransmitter gamma aminobutyric acid (GABA) to reduce the release of the excitatory neurotransmitter glutamate, thereby preventing the spread of the seizure within the brain. Current antiepileptic drugs are effective in approximately 50 percent of patients. However, 30 percent of the epileptic patient population experience intractable seizures regardless of the antiepileptic drug used, and 50 percent of the population will eventually become resistant to currently available treatments. All existing antiepileptic drugs are associated with numerous side effects (impairment of motor function, cognitive dysfunction, emotional lability). Therefore, there is a need for the development of better treatment options (Williams, Jones, and Whalley 2014). Indeed, pure oral CBD (Epidiolex) has recently been approved by the FDA as a treatment for rare forms of childhood epilepsy.

## Historical Aspects of CBD and Epilepsy

Cannabis has been used as a medicinal plant since ancient times. There are indications of some use as an antiepileptic agent in India and possibly in Assyria (Russo 2017), but neither the ancient Greeks nor the Romans and most of the nations in the ancient Middle East (up to the first centuries AD) seem to have used it widely for this medical condition (Mechoulam 1986). However, it was well known later to Arab physicians. Thus, Ibn al Badri, in a treatise on hashish written around 1464 (preserved in Paris in a manuscript form), tells that the poet Ali ben Makki visited the epileptic Zahir-ad-din Muhammed, the son of the Chamberlain of the Caliphate Council in Baghdad, and gave the reluctant Zahir-ad-din hashish as medication. It cured him completely, but he could not be without the drug ever after (Rosenthal 1972).

During the nineteenth century, several medical reports were published on the ameliorative effects of cannabis extracts in several forms of convulsions. In 1839, William O'Shaughnessy (1839) described the successful treatment of seizures in an infant by using cannabis tincture. Queen Victoria's personal physician, J. R. Reynolds, described cannabis as the most useful agent that he was acquainted with to treat violent convulsions (Reynolds 1868). We now know that whole cannabis contains several cannabinoids with diverse pharmacology, so it is difficult to interpret the mechanisms of action of whole cannabis plant use (Williams, Jones, and Whalley 2014).

## Preclinical Studies: CBD and Epilepsy

CBD is the only isolated phytocannabinoid to have been investigated for anticonvulsant effects both preclinically in animals and clinically in humans. The first preclinical evidence (Izquierdo, Orsingher, and Berardi 1973; Karler, Cely, and Turkanis 1973) showed that injected CBD (1.5–12 mg/kg, i.p.) and oral CBD (120 mg/kg) reduced seizures in mice. Karler and Turkanis subsequently showed that CBD (0.3–3 mg/kg, i.p.) increased the threshold to induce limbic seizures in rats, as did the antiepileptic drug phenytoin; however, CBD went beyond phenytoin by also reducing aspects of seizure recordings such as the after-discharge amplitude, duration, and propagation, all hallmarks of seizure recodings (Turkanis et al. 1979). The authors concluded that CBD was the most efficacious of the drugs tested against limbic after-discharges and convulsions. It is of interest that CBD and phenytoin chemical structures share a similar spatial relationship between the two rings and have close crystal structures. Hence, CBD and phenytoin fulfill the stereochemical requirements suggested for anticonvulsant drug action (Tamir, Mechoulam, and Meyer 1980), with CBD being more efficacious.

Karler and Turkanis (1980) also compared tolerance development to the anticonvulsant activity of THC with that of CBD in three electrically induced seizure-threshold tests. Although tolerance developed to THC in some of the tests, no tolerance was observed with CBD.

CBD clearly shows anticonvulsant effects in acute preclinical models of epilepsy, but the mechanism of action is not well understood. CBD may reduce neuronal excitability and neural transmission by reducing intracellular calcium through interaction with TRPV1 receptors or GPR55 receptors (Devinsky et al. 2014). GPR55 receptors are located on excitatory axon terminals, where they facilitate glutamate release when the neuron fires. Because CBD effectively blocks GPR55 activation, it would be an ideal candidate as an anticonvulsant agent by selectively dampening excess presynaptic glutamate release from only the hyperactive excitatory neurons during epileptic seizures (Katona 2015).

CBD may not be as effective in animal models of chronic epilepsy. Cortical implantation of cobalt was used to model chronic seizures in humans (Colasanti, Lindamood, and Craig 1982); under these conditions, CBD (60 mg/kg, i.p.) was ineffective. Recent work has verified that CBD shows significant anti-epileptiform and anticonvulsant activity in a wide variety of in vitro and in vivo models (Jones et al. 2010). Cannabidivarin, the propyl variant of CBD, has recently been shown to have anticonvulsant properties in the same models (Hill et al. 2012).

These preclinical findings set the stage for the human clinical trial testing of CBD specifically for the treatment of epilepsy. In fact, preclinical studies found that CBD was a more reliable anticonvulsant than THC, and without THC's psychoactive or motor side effects. Therefore, CBD appears to be a better therapeutic option for epilepsy than THC.

## Human Clinical Trials of CBD in Epilepsy

Over forty years ago, on the basis of preclinical animal research indicating the effectiveness of CBD in reducing epileptic seizures, Mechoulam and Carlini (1978) provided the first scientific clinical trial of the potential of CBD to reduce epileptic seizures. In a small-scale, double-blind, placebo-controlled trial, patients received either 200 mg CBD daily (four patients) or placebo (five patients) for three months, in addition to their normal antiepilepsy medication. Of the patients in the CBD group, two had no seizures for the entire three-month period, one partially improved, and the fourth had no improvement. For the patients in the placebo group, no improvements were observed. No toxic effects were reported for either group. These results are very promising, but this study was limited by the small samples. In a subsequent study (Cunha et al. 1980), fifteen patients with "secondarily generalized epilepsy with temporal focus" were randomly divided to receive 200 to 300 mg daily CBD ($n=8$) or placebo ($n=7$) for up to four and a half months in combination with their prescribed medications. This group of patients was selected because the current combined medications were no longer effective in the control of their symptoms. CBD completely prevented seizure episodes in four of the eight patients, partially improved clinical symptoms in three patients, and was ineffective in one patient. Placebo was ineffective in seven patients, with one showing some improvement. The patients reported no severe side effects of treatment. Despite these initial promising results, which were consistent with the preclinical animal literature, it is only very recently that subsequent large-scale, double-blind, placebo-controlled clinical trials have been conducted, mostly in response to the large number of anecdotal reports of the positive effects of CBD-rich marijuana on children with pediatric epilepsy. Why did we have to wait for decades (Mechoulam et al. 2014)?

Even though the efficacy of CBD for treating epilepsy in humans was supported nearly forty years ago, until recently the published data on the use of CBD for the treatment of epilepsy came from fewer than seventy participants, very few of them children (Ames and Cridland 1986). Few of the studies were rigorous, and few of them used high-quality evidence (Whiting et al. 2015). More recently, Porter and Jacobson (2013) published self-reports of the experiences of parents who had given their children (nineteen in all) some form of high-CBD product for severe intractable epilepsy. The reported doses of CBD ranged from below 0.5 mg/kg/day to about 30 mg/kg/day. The majority of the families reported improvement, ten of them reporting greater than 80 percent improvement and two reporting complete cessation of seizures. Others reported having discontinued other medications. Although this does not constitute a high-quality placebo-controlled experiment, it does provide some information that was missing until 2013. A number of families in the United States and in Canada have gained access to various "hemp oil" preparations with high CBD and low THC content. Assessing the outcomes of the use of such preparations is difficult because of the high variability among products, lack of consistency in dosing, variable quality control, and uncertainty about the presence of other potentially active cannabinoids. In a report from Colorado, although 57 percent of families using such products reported positive results, there was no evidence of improvement in the EEG pattern in eight of the patients who reportedly responded to the drug. And there were significant adverse effects, including increased seizures in 13 percent of the patients (Press, Knupp, and Chapman 2015). Interestingly, a higher rate of benefit was reported by families that had moved to Colorado specifically in order to gain access to "hemp oil" products than by families that had already been living there, which suggests a placebo effect in self-reported outcomes.

It is possible that some of the adverse effects of CBD in these studies may relate to interactions with other antiepileptic drugs. In a study on thirteen patients with refractory epilepsy concomitantly taking clobazam (a common antiepileptic drug), it was found that nine of thirteen patients had a greater than 50 percent decrease in seizures. Side effects were reported in ten of the patients (77 percent) but were alleviated with clobazam dose reduction (Geffrey et al. 2015). In fact, recent studies using pharmacokinetic modeling show that an interaction with concomitantly used clobazam is a plausible explanation for CBDs "therapeutic" effects in Lennox–Gastaut and Dravet syndrome (Balachandran et al., 2021; Bergmann, Broekhuizen and Groeneveld, 2020).

### The Development of CBD (Epidiolex), an FDA-Approved Treatment for Treatment-Resistant Childhood Epilepsy

The treatment of a highly resistant, rare form of epilepsy in children, Dravet syndrome, which has a very high mortality rate, is the indication for which there is the strongest clinical evidence of efficacy of CBD. Beginning in the second year of life, children affected by

Dravet syndrome develop an epileptic encephalopathy that results in cognitive, behavioral, and motor impairment (Devinsky et al. 2014). Research on the treatment of this condition was accelerated by widespread press coverage of anecdotal successes in 2016. Sanjay Gupta, chief medical correspondent for the television channel CNN, has produced two hour-long specials on the efficacy of CBD-rich cannabis in treating Dravet syndrome, which is usually resistant to standard antiepilepsy drugs.

One of the first studies, Devinsky et al. (2016), was an open label study of 214 patients (aged 1–30 years) receiving stable doses of antiepileptic drugs before the study. There was no placebo control group. All patients were initially given oral doses of CBD of 2–5 mg/kg/day and then titrated until intolerance or to a maximum dose of 25–50 mg/kg/day depending on study site. The main measure was percentage change in frequency of seizures. The mean monthly frequency of motor seizures was reduced from 30 at baseline to 15.8 over the twelve-week treatment period. Although the trial was primarily aimed to assess safety, the absence of a placebo control group means that the results cannot be used to assess the likelihood that CBD itself produced particular effects.

In a subsequent article in the *New England Journal of Medicine*, Devinsky et al. (2017) reported the results of a controlled trial of CBD treatment for Dravet syndrome. In a double-blind, placebo-controlled trial, 120 children and young adults with Dravet syndrome were randomly assigned to receive either CBD oral solution (20 mg/kg/day) or placebo in addition to standard antiepileptic treatment. CBD reduced the median frequency of convulsive seizures per month from 12.4 to 5.9 (with 5 percent of the sample becoming seizure free) as compared with a decrease from 14.9 to 14.1 with placebo. Adverse effects, occurring more frequently in the CBD group than the placebo group, included diarrhea (31 percent versus 10 percent), loss of appetite (28 percent versus 5 percent), and somnolence (36 percent versus 10 percent). Adverse effects led to the withdrawal of eight patients in the CBD group and only one patient in the placebo group.

A similar article appearing in *Lancet* (Thiele et al. 2018) evaluated the efficacy and safety of CBD in patients with Lennox-Gastaut syndrome, epileptic encephalopathy that produces various types of treatment-resistant seizures, including drop seizures. This was a multisite study with twenty-four clinical sites located in the United States, the Netherlands, and Poland. A total of 171 patients (ages 2–55 years old) were randomized to receive CBD (200 mg/kg, oral solution, with $n=86$) or matched placebo ($n=85$) while they continued to receive their typical antiepileptic regimen (an average of three medications per patient in each group). Treatments were given daily for fourteen weeks, with two weeks of dose escalation (beginning at 2.5 mg/kg of CBD) and twelve weeks of maintenance at the 200 mg/kg dose of CBD. At the end of treatment, a ten-day dose taper was included. CBD treatment decreased drop seizure frequency by a median of 43.9 percent (71.4 seizures per patient per month at baseline reduced to 31.4 during treatment), compared to a 21.8 percent reduction in the placebo group (74.7 at baseline, 56.3 during treatment). As well, the number of patients experiencing greater than or equal to 50 percent reduction in drop seizure rate was increased by CBD (44 percent, $n=38$) relative to placebo (24 percent, $n=20$). Other types of seizures were also reduced in the CBD group (49.4 percent reduction in the CBD group, 22.9 percent reduction in the placebo group). However, treatment-related adverse events occurred more frequently in the CBD group than the placebo group: diarrhea (13 percent versus 4 percent), somnolence (14 percent versus 8 percent), decreased appetite (9 percent versus 1 percent), and vomiting (7 percent versus 5 percent). Increases in liver function tests (greater than three times the upper limit of normal) occurred in twenty patients in the CBD group and one patient in the placebo group.

These successful trials resulted in the approval of a pure oral CBD product (Epidiolex) for these conditions by the FDA in the United

States in 2018, and it is now marketed in North America and in Europe for Dravet syndrome and Lennox-Gastaut syndrome. Recent trials (Herlopian et al. 2020; Miller et al. 2020) have given support to this clinical use.

## Conclusion

The strongest clinical evidence for the efficacy of CBD as an adjunct therapeutic agent is in the treatment of rare childhood epilepsy. Epidiolex is an FDA-approved medical treatment for these disorders. Further clinical trials are essential to determine if CBD is also an effective treatment for epilepsy in adulthood. Indeed, there are forty current clinical trials indicated on the NIH website, www.clinicaltrials.gov. Approximately half of these studies include participants with childhood epilepsy, and the others are across ages. The hope is that the results of these trials will provide an answer about the generality of CBD as a treatment for epilepsy and perhaps as a treatment for nonepileptic seizures.

# 5
# Neuroprotection, Tissue Protection, and Cancer

Several in vitro and in vivo preclinical reports have demonstrated neuroprotective, tissue-protective, and even tumor-reduction effects of CBD in rodent models (Pacher, Kogan, and Mechoulam 2020). Unfortunately, little has been done to translate these preclinical observations into clinical trials to evaluate their efficacy in human patients. In this chapter, we review the existing literature, which is primarily based on results from in vitro and preclinical animal studies.

## CBD as a Redox Modulator

Antioxidants are chemicals that interact with and neutralize free radicals; they are functionally "free radical scavengers." Free radicals (created when an atom or a molecule gains or loses an electron) are formed naturally in the body and are important for many normal cellular processes. At high concentrations, however, free radicals are hazardous to the body and damage components of cells, including DNA, proteins, and cell membranes. The damage to cells caused by free radicals, especially damage to DNA, may play a role in the development of cancer and other health conditions.

The most common type of free radicals produced in living tissue are those that contain oxygen, called reactive oxygen species (ROS). Antioxidants reduce ROS produced by toxic events, such as cerebral ischemia (stroke). The research group of Julius Axelrod (Hampson et al. 1998) showed that both THC and CBD are more potent neuroprotective antioxidants than the classic reference antioxidants tested at the time. These antioxidant effects of CBD are responsible for many of the neuroprotective effects discussed in this chapter. CBD appears to play a role as a redox modulator to maintain healthy cells (Singer et al. 2015); it protects healthy cells from damage by ROS, but it may also promote the death of proliferating unhealthy cancer cells by activation of ROS, among other potential mechanisms. Singer and colleagues (2015) have shown that combining CBD treatment with inhibition of the natural antioxidant response system by Xc (catalytic subunit SLC7A11) results in ROS increase due to synergism, leading to robust antitumor effects. It is possible that some of the effects of CBD in neurological diseases may also involve ROS increase due to such synergistic effects.

Oxidative stress is involved in the pathophysiology of numerous diseases of the nervous system. A recent review of this literature (di Giacomo et al. 2020) indicates that neurons are particularly susceptible to redox changes due to their high metabolic rate and limited antioxidant capacity. Astrocytes have greater antioxidative potential than neurons; astrocytic support of the neuronal antioxidant system is a key neuroprotective mechanism against oxidative damage. Although there is still a limited understanding of the mechanisms of CBD neuroprotection, it is clear that the drug's effects are not limited to the direct modulation of neurons but are extended to microglia, oligodendrocytes, and astrocytes (Scarante et al. 2020). In different models of brain ischemia and neurogenesis, the positive effects exerted by CBD are strongly related to the astrocyte response to the compound (Ceprian et al. 2019; Lafuente et al. 2011; di Giacomo et al. 2020).

## CBD and the Immune System

The immune response is a delicate balance between robust reactions to defend the body against foreign cells but limited or no reaction to self-cells (Nichols and Kaplan 2020). Various cell types act together to provide protection against foreign invaders but also to avoid reactions against self-proteins. The innate immune system consists of neutrophils, macrophages, and other myeloid cells, which readily destroy pathogens. If the innate response is insufficient, then these cells can activate the adaptive immune response consisting predominantly of T and B cells. T cells (thymus cells) and B cells (bone marrow- or bursa-derived cells) are the major cellular components of the adaptive immune response. T cells are involved in cell-mediated immunity, whereas B cells are primarily responsible for humoral immunity (relating to antibodies).

T cells can send signals that recruit and activate other immune cells or directly induce apoptosis of infected cells. T cells also help stimulate B cells, which produce antibodies to neutralize pathogens or enhance destruction of pathogens. The innate immune response and T cell activation promote inflammation as a result of pathogen destruction–induced tissue damage, which results in the production of proinflammatory cytokines. Typical proinflammatory cytokines include IL-1α, IL-1β, IL-6, TNF-α, and IL-17A. Some cytokines are produced by specific T cell subsets; for instance, the Th1 subset produces interferon-gamma (IFN-γ), which promotes cell-mediated cytotoxicity, while the Th2 subset produces IL-4 and promotes β cell responses (Nichols and Kaplan 2020). Other end points that can provide clues of disruption of immune responsiveness are nitric oxide or myeloperoxidase (MPO) produced from innate cells during pathogen destruction.

The effects of CBD on immune responses can involve innate or adaptive responses. CBD has consistently been shown to reduce cytokine production, produce apoptosis, and suppress nitric oxide in vitro (Nichols and Kaplan 2020). The effect of CBD on nitric oxide is mediated by suppression of inducible nitric oxide synthase (iNOS) in response to inflammatory stimulation. iNOS is regulated by the transcription factor nuclear factor-κB (NF-κB), which CBD has been shown to suppress. CBD treatment in vivo in preclinical models has been shown to lower IL-6 production in peritoneal macrophages stimulated with lipopolysaccharide (LPS) (Weiss et al. 2008). CBD also suppresses neutrophils by decreasing their numbers and compromising MPO activity. The area in which most of the effects of CBD in the immune system have been studied is T cells; CBD inhibits the production of IFN-γ in T cells in vitro and in vivo (Nichols and Kaplan 2020). Many of the mechanisms of CBD on innate immune cells also account for CBD's ability to decrease microglial cell activation and produce apoptosis of microglial cells (Nichols and Kaplan 2020). In vivo, CBD decreases microglial accumulation in the spinal cord in diabetic mice, which might contribute to its ability to attenuate neuropathic pain (Toth et al. 2010). There is a planned, randomized, open-label interventional study assessing CBD and THC on immune cell activation in HIV patients (Costiniuk et al. 2019), which will be the first human trial to assess the tolerability of CBD in human HIV patients, as well as the effect of CBD on inflammatory markers in these patients.

## CBD and Viral Diseases

Coronavirus disease 2019 (COVID-19), caused by the betacoronavirus severe acute respiratory syndrome coronavirus 2 (SARS-CoV-2), is a respiratory disease currently causing a global pandemic. Growing unsubstantiated claims on the Internet suggest that CBD could be used for treatment of COVID-19 in human patients, which have triggered FDA reprimand letters (Shover and Humphreys 2020). Because CBD shows anti-inflammatory activity and is approved by the FDA for seizure reduction treatment of children with intractible epilepsy, a number of reviews have suggested that CBD might alleviate COVID-19 symptoms, at least during the later stages of the disease during a phase called the cytokine storm (see Hill 2020; Brown 2020; Costiniuk and Jenabian 2020; Mamber et al. 2020). Coronaviruses represent a large family of viruses, but CBD's effects on this virus family (or any other) have not been investigated in human clinical trials. (For an excellent review of the advantages and disadvantages of CBD as a potential agent for the treatment of COVID-19 on the basis of results from in vitro and preclinical studies, see Malinowska et al. 2021.)

In vitro evidence suggests that CBD may reduce entry of SARS-CoV-2 into cells by downregulating the proteins (angeotensin-converting enzyme 2; ACE2 and transmembrane serine protease 2:TMPRSS2) responsible for viral entry (Wang et al. 2020). As well, a direct antiviral effect of CBD has also been demonstrated in cultured Vero cells infected with SARS-CoV-2 (Raj et al. 2021); CBD exhibited an $IC_{50}$ value of 8 μM for its inhibitory effect on SARS-CoV-2 replication and was at least as potent as remdesivir, another potential antiviral compound being developed for the treatment of COVID-19.

There is recent in vitro and limited in vivo support for the effectiveness of CBD to inhibit SARS-CoV-2 replication in cells (Nguyen et al. 2022). CBD reduced replication in human lung epithelium cells by inhibiting viral gene expression after viral entry into the cells. In addition, CBD reduced viral titers in the lung and nasal turbinates of SARS-CoV-2 -infected mice, providing the first preclinical evidence in support of CBD as a potential treatment for this viral disease (Nguyen et al. 2022). CBDA and CBGA have also been shown in vitro to prevent entry of live SARS-CoV-2 into human epithelial cells (Van Breemen et al. 2022). CBD has previously been shown in vitro to inhibit the replication of hepatitis C virus but not hepatitis B virus (Lowe, Toyang, and McLaughlin 2017). Finally, a preclinical mouse study has demonstrated the effectiveness of CBD to reduce the acute respiratory distress syndrome and cytokine storm induced in mice by poly (I:C), a synthetic analogue of viral double-stranded RNA (Khodadadi et al. 2020), presumably acting to reduce inflammation. The database at www.clinicaltrials.gov lists eight clinical trials in which the use of CBD is being assessed in the context of COVID-19.

Although some COVID-19 patients show only mild fever, cough, or muscle soreness, some patients suddenly deteriorate in the later stages of the disease. In this later stage of disease, SARS-CoV-2 can trigger a cytokine storm, producing excessive levels of pro-inflammatory mediators, resulting in a hyperinflammation condition (Mehta et al. 2020). Therefore, treatments that can reduce this late cytokine storm phase of SARS-CoV-2 associated inflammation, such as the glucocorticoid dexamethasone, have been reported recently in the *New England Journal of Medicine* (2021) to be beneficial to patients receiving invasive ventilation or oxygen alone (RECOVERY Collaborative Group 2021). The authors note, however, "that the beneficial effects of glucocorticoids in severe viral respiratory infections are dependent on the selection of the right dose, at the right time, in the right patient. High doses may be more harmful than helpful, if such treatment is given at a time when control of viral replications is paramount and inflammation is minimal."

Because of its anti-inflammatory effects, several reviews have suggested that CBD may also be effective in reducing the cytokine storm in these patients. A recent retrospective study of 150 confirmed COVID-19 cases in Wuhan, China, suggests that elevated inflammatory indicators such as IL-6 were predictors of fatality (Ruan et al. 2020). Severe COVID-19 cases may benefit from IL-6 pathway inhibition (Fu, Xu, and Wei 2020), and indeed tocilizumab, a treatment that blocks the IL-6 receptor, has shown some clinical promise (Xu, Han et al. 2020). Interestingly, in rodent models of asthma and acute lung injury, CBD treatment also reduced serum levels of cytokines, including IL-6, suggesting that CBD controls the exaggerated inflammatory response observed in these animal models (Vuolo et al. 2015; Ribeiro et al. 2012, 2015). Most recently, Anil et al. (2021) showed that CBD exhibited anti-inflammatory activity in lung epithelial cells by reducing IL-6 and IL-8 (two of the prominent cytokines that characterize the cyokine storm in patients with severe COVID-19) in an alveolar cell line. However the anti-inflammatory effects of CBD were seen only within a very narrow dose range, as has been previously demonstrated, limiting potential usefulness in clinical therapy. Whether such effects would also be seen in human patients is unknown.

Although the use of CBD as an anti-inflammatory treatment seems logical, reducing the activity of inflammatory mediators may not actually be an advantage when dealing with viruses because it may mitigate host immune responses to acute viral infections, which could lead to disease progression and even death (Reiss 2010; Tahamtan et al. 2016). The timing of treatment administration may be critical. As reviewed by Brown (2020), in a phase 3 clinical trial of Epidiolex for seizure disorders, relative to placebo, infections were 30 percent more common in CBD patients. Among infections, viral (11 percent CBD versus 6 percent placebo) and pneumonia (5 percent CBD versus 1 percent placebo) had the largest increases with CBD treatment. This suggests that CBD may actually dampen the ability to fight off infections, which may not be a therapeutic advantage.

In addition to its anti-inflammatory effects, CBD can also induce apoptosis in mammalian cells, thought to be an essential component of host responses to viral infections (Sledzinski et al. 2018; Orzalli and Kagan 2017). The current state of evidence for this claim was recently evaluated (Tagne et al. 2020).

As scientists learn more about this novel SARS-CoV-2, better treatment strategies may be elucidated. Indeed, a focus on the asymptomatic COVID-19 patients may be the best source of information as they have been infected with the virus, but their body was able to defend itself. What is different physiologically about these patients?

**CBD and Cerebral Ischemia**

During a cerebral ischemic episode, commonly known as a stroke, large quantities of the excitatory neurotransmitter glutamate are released, causing neuronal death by overstimulating the glutamate receptors. This overstimulation results in metabolic stress and accumulation of toxic levels of calcium in the cell. In vitro and in vivo studies have shown that this neurotoxicity can be reduced by antioxidants, as well as glutamate receptor antagonists. Antioxidants are neuroprotective because of their ability to reduce toxic ROS formed at the time of ischemic stress. CBD ($ED_{50}$ 2–4 µM) was found to potently prevent cell death produced by glutamate in rat cortical neuron cultures through its antioxidant effect (Hampson et al. 1998, 2000).

CBD has also been shown to be neuroprotective against ischemia in an in vivo rat stroke model by acting as an antioxidant (Hampson et al. 2000). In anesthetized rats, a suture was fed through the carotid artery up into the middle cerebral artery (MCA). The suture prevented blood flow and was left in place for ninety minutes and then removed. The animals recovered for a forty-eight-hour period before a battery of neurological tests was administered. After these tests, the rats were sacrificed, their brains were fixed and sliced, and the area of infarct was calculated by computer imaging. Prior to the ischemia, rats were intravenously (i.v.) administered 5 mg/kg of CBD or vehicle (the solution that CBD is mixed in, used as a control). A second 20 mg/kg dose of CBD or vehicle was administered intraperitoneally (i.p.) twelve hours after surgery. Examination of the computer brain images revealed that the infarct size produced by the loss of blood flow was reduced by 60 percent in the CBD-treated rats compared with the vehicle-treated rats. As well, the neurological tests revealed that the CBD-treated rats were significantly less impaired in neurological tests than the vehicle-treated rats. Subsequent research with gerbils has shown that CBD (1.25–20 mg/kg, i.p.) given five minutes after the ischemic event produced a dose-dependent, bell-shaped curve for neuroprotective effects with the greatest efficacy at 5 mg/kg (Braida et al. 2003). Furthermore, CBD given immediately before and three hours after an ischemic event has also been shown to reverse the brain damage in mice by acting as an antioxidant and as a 5-$HT_{1A}$ receptor agonist (Mishima et al. 2005); tolerance did not develop to the neuroprotective effects of

CBD over fourteen days of treatment (Hayakawa et al. 2008). There have been no human trials demonstrating the potential of CBD to restore function in cases of cerebral ischemia.

## CBD and Cardiovascular Effects

The neuroprotective effects of CBD have been recently reviewed (Pacher, Kogan, and Mechoulam 2020). Myocardial infarction, commonly known as a heart attack, occurs when the heart receives insufficient blood flow, causing necrosis (heart tissue damage). CBD has been shown to decrease myocardial necrosis in rat and rabbit models of myocardial infarction by attenuating oxidative stress and improving antioxidant defense mechanisms (Durst et al. 2007; Feng et al. 2015). CBD has also been shown to have beneficial effects in animal models of cardiomyopathy, a condition causing the heart muscle to become enlarged. In rat and mouse cardiomyopathy models produced by the toxic chemotherapeutic drug doxorubicin, CBD improved cardiac function and attenuated apoptosis of cells by decreasing oxidative stress (Hao et al. 2015; Fouad, Al-Mulhim, and Gomaa 2013). Chronic CBD treatment also improved type I diabetes–induced myocardial dysfunction and cell death by attenuating NFκB (nuclear factor kappa enhancer of activated B cells) activation, a protein complex that controls transcription of DNA, cytokine production, and cell survival (Rajesh et al. 2010; Pacher, Kogan, and Mechoulam 2020). These preclinical findings provided the proof of principal rationale for an exploratory clinical trial to assess the safety of CBD in patients with heart failure (NCT03634189).

## CBD and Liver and Kidney Injury and Disease

CBD has been shown in rodents to reduce hepatic ischemia reperfusion and alcohol binge-induced liver injury (Mukhopadhyay et al.

2011). CBD reduced the extent of liver inflammation and oxidative stress and cell death by attenuating key inflammatory markers. As well, it improved cognitive function in a mouse model of hepatic encephalopathy produced by bile duct ligation (Magen et al. 2009). A nine-week (eight-week treatment period and one-week safety follow-up) randomized partially blinded study evaluated the effects of CBD on human participants with raised liver triglycerides (liver fat 5 percent or more) clinically diagnosed with fatty liver disease (www.clinicaltrials.gov, NCT01284634: GW Pharmaceuticals). Doses of CBD ranging from 200 mg/day to 800 mg/day reduced liver triglyceride levels from baseline among the CBD-treated patients but not among the placebo-treated patients.

## CBD and Inflammation and Colitis

The anti-inflammatory effects of CBD shown in most models are mediated by the attenuation of T cell infiltration or proliferation, microglial activation, and consequent oxidative stress and inflammatory response (Pacher, Kogan, and Mechoulam 2020). Promising preclinical data led to a human clinical trial in patients with ulcerative colitis with a CBD-rich botanical extract. Although the primary end point (the percentage of patients in remission after treatment) was not reached (CBD 28 percent, placebo 26 percent), CBD did reduce some symptoms of ulcerative colitis (Irving et al. 2018). (See also chapter 7 for the effects of CBD on gastrointestinal tract inflammation and irritable bowel disease.)

## CBD and Autoimmune Diseases

An abnormal immune response/attack on normal body cells may affect multiple organ systems and lead to an autoimmune disease.

There are about eighty known autoimmune diseases. The mechanism of most of these diseases has not been fully clarified. It is widely assumed that some autoimmune diseases are triggered by infections, which lead to the production of antibodies that may not be completely specific for the infectious agent and may interact with other proteins. Environmental factors and genetics are also involved. CBD has been shown to affect some autoimmune diseases. (For a recent review, see Nichols and Kaplan, 2020.) Below we present the published data on some autoimmune (or autoimmune-like) diseases. Published data on psoriasis, multiple sclerosis and arthritis are presented in other parts of this book.

## Diabetes Type 1

In this disease, the immune system destructs the insulin-producing islets-of-Langerhans beta cells in the pancreas. Weiss and colleagues (2006) reported that CBD lowered the incidence of diabetes in young, nonobese diabetes-prone (NOD) female mice. It also significantly reduced the plasma levels of the proinflammatory cytokines IFN-γ and TNF-α, lowered Th1-associated cytokine production in vitro, and augmented Th2-associated cytokines IL-4 and IL-10. In a later publication, the same group (Weiss et al. 2008) showed that administration of CBD (five injections per week for four weeks) to eleven- to fourteen-week-old female NOD mice, which were either in a latent diabetes stage or with initial symptoms of diabetes, ameliorated the manifestations of the disease. Diabetes was diagnosed in only 32 percent of the mice in the CBD-treated group, compared to 86 percent and 100 percent in the vehicle-treated and untreated groups, respectively. In addition, the level of the proinflammatory cytokine IL-12 produced by splenocytes was significantly reduced, whereas the level of the anti-inflammatory IL-10 was significantly elevated following CBD treatment. Histological examination of the pancreas cells of CBD-treated mice revealed more intact islets than in the controls. The authors suggest that CBD, known to be safe in humans, can possibly be used as a therapeutic agent for treatment of type 1 diabetes.

The positive aspects of these observations are supported by data, obtained by Lehmann et al. (2016), who used experimental methods different from those of the Weiss et al. group. Lehmann et al. administered daily 5 mg/kg CBD for ten weeks to seven-week-old female NOD mice. Leukocyte activation and functional capillary density were quantified by intravital microscopy. They found that CBD treatment reduced early pancreatic inflammation in type 1 diabetes. In view of the positive results in mice by two different groups using different methods, CBD should have been tested in human patients. However, no clinical work has been reported. Again, are we missing something?

CBD was also found to improve depression-like behavior and anxiety-like behavior in mice with diabetes type 1 (Chaves et al. 2020). This observation may be of relevance to human diabetes type 1 patients in whom depression and anxiety are at least two times higher than in healthy individuals.

**Graft versus Host Disease (GvHD)**

Bone marrow transplants are important medical treatments in certain cancer conditions. However the donated white blood cells, which remain within the graft, view the host as a foreign tissue and may attack the recipient's body cells, causing GvHD. This effect can be considered a reverse autoimmune reaction and is a major obstacle to successful allogeneic hematopoietic cell transplantation (alloHCT).

As CBD is a potent anti-inflammatory and immunosuppressive agent, it was assumed that it may decrease GvHD incidence and severity after alloHCT (Yeshurun et al. 2015). In a clinical trial, forty-eight adult patients undergoing alloHCT were enrolled. GvHD prophylaxis consisted of cyclosporine and a short course of methotrexate; patients, who received the transplant from an unrelated donor, were also administered low-dose anti-T cell globulin. CBD

(300 mg/day) was given orally starting seven days before transplantation until day 30. The median follow-up was sixteen months. None of the patients developed acute GvHD while consuming CBD. A comparison between patients who received CBD and patients who did not was reported. Patients who received CBD developed significantly less GvHD (including very severe disease). The authors conclude that the combination of CBD with standard GvHD prophylaxis is a safe and promising strategy to reduce the incidence of acute GvHD. A pharmaceutical company (Kalytera Therapeutics) has published that it is developing CBD treatment for GvHD.

## Experimental Autoimmune Hepatitis

Administration of CBD inhibited experimental hepatitis (Hegde, Nagarkatti, and Nagarkatti 2011), as well as all of the associated inflammation markers. Acute hepatitis was caused by injection of concanavalin A (ConA) in C57BL/6 mice. It was characterized by a significant increase in aspartate transaminase, induction of inflammatory cytokines, and infiltration of mononuclear cells in the liver. The administration of CBD was made after that of ConA. The mechanism was shown to involve CBD induction of myeloid-derived suppressor cells (MDSCs), which suppressed T cell proliferation and thus conferred significant protection from ConA-induced hepatitis. It was further shown that MDSC suppressed hepatitis through the vanilloid receptor TRPV1. Mice deficient in TRPV1 were not protected. The authors conclude that "CBD, which triggers MDSCs through activation of TRPV1 vanilloid receptors, may constitute a novel therapeutic modality to treat inflammatory diseases."

## Autoimmune Myocarditis

Lee and colleagues (2016) have reported the results of a study of the effects of CBD on a well-established mouse model of experimental autoimmune myocarditis (EAM), an autoimmune disease that is characterized by inflammation of the heart. This heart disease was induced by immunization with cardiac myosin, which led to T cell–mediated inflammation, cardiomyocyte cell death, fibrosis, and myocardial dysfunction. EAM was characterized by marked myocardial T cell infiltration, profound inflammatory response, and fibrosis accompanied by marked attenuation of both systolic and diastolic cardiac functions. Chronic treatment with CBD largely attenuated the $CD3^+$ and $CD4^+$ T cell–mediated inflammatory response and injury, myocardial fibrosis, and cardiac dysfunction in mice. The authors conclude that "CBD may represent a promising novel treatment for managing autoimmune myocarditis and possibly other autoimmune disorders and organ transplantation."

## Various Autoimmune Disease States

**Crohn's Disease** Patients using cannabis in various forms have reported that it has a positive effect in their disease management. However, the amount of CBD in medical cannabis varies, so conclusions cannot be drawn. Naftali and colleagues (2017) showed that administration of pure oral CBD (10 mg, twice daily for eight weeks) to Crohn's patients had negative results. However, the CBD dose administered was rather low. (For a review, see Naftali 2020. See also chapter 7 for a discussion of CBD's effects in irritable bowel syndrome.)

**Systemic Lupus Erythematosus** Systemic lupus erythematosus (SLE) is a chronic autoimmune disease that causes inflammation and tissue damage in organs such as the skin, brain, lungs, and kidneys. The effect of CBD (25 mg/kg, s.c. daily beginning at fourteen weeks of age) on disease progression in a murine model of SLE was evaluated over twenty weeks (Katz-Talmor et al. 2018). CBD caused progression of glomerular disease with significantly increased proteinuria. A trend toward lower survival was noted. The authors' conclusion was that CBD accelerated disease progression. This observation has not been made in either murine or clinical trials on other diseases and seems to be specific for SLE.

## CBD and Neurodegenerative Diseases

Neurodegenerative diseases affecting the CNS, including multiple sclerosis (MS), Alzheimer's disease (AD), Parkinson's disease (PD), and Huntington's disease (HD), are characterized by progressive loss of specific neuronal subpopulations affecting selective regions of the brain, spinal cord, and neurotransmitter systems that produce different clinical symptoms. Currently, there is no cure for any of these disorders. Cannabinoids may have promise in symptom alleviation or slowing disease progression for these disorders, supported by in vitro and in vivo evidence; however, the evidence that CBD may be effective in treating these diseases is limited.

### Multiple Sclerosis (MS)

MS is an inflammatory disease of the brain and spinal cord in which infiltration of lymphocytes produces damage to myelin and axons. During the early phase of the disease, inflammation is only temporary and remyelination occurs, resulting in a recovery episode for patients. However, over time, widespread microglial activation produces chronic neurodegeneration that results in the patient's progressive increasing disability. The medications currently available for MS can reduce the frequency of new episodes but cannot reverse the course of the disease (Pryce and Baker 2014).

Cannabis has become of interest in controlling some of the symptoms of MS including bladder incontinence, tremor, and spasticity (Pryce and Baker 2014); indeed, nabiximols (Sativex [2.7 mg THC: 2.5 mg CBD/spray]) has been approved in Europe and Canada for treatment of MS spasticity and neuropathic pain. Several controlled trials to assess the safety and efficacy of nabiximols on symptoms of MS were conducted over a decade ago (Friedman, French, and Maccarrone 2019). Nabiximols versus placebo were compared to treat several symptoms of MS (spasticity, spasms, bladder problems, tremor, and pain) in 160 randomized patients with MS. There was no overall difference in composite symptom score, but 37 patients indicated improvement in spasticity (Wade et al. 2004). A second trial with 337 patients with moderate to severe spasticity reported symptom reduction (patient reported numerical rating scale) with nabiximols compared with placebo (Collin et al. 2010). The American Academy of Neurology evidence-based review concluded that nabiximols was probably effective in reducing patient-reported symptoms of spasticity but not objective measures of spasticity (Novotna et al. 2011). In a third human clinical trial for MS symptoms, patients entered a four-week phase A treatment to identify responders to nabiximols treatment (approximately half of the participants), who were randomly assigned to receive nabiximols or placebo in phase B. In phase B, patients receiving nabiximols had a 30 percent or greater reduction in spasticity compared to those receiving placebo (Novotna et al. 2011; Markova et al. 2019). Most recently in a fourth trial, 15 patients with progressive MS were compared with 14 healthy controls before and during a standard treatment with nabiximols using clinical scales for spasticity and pain, as well as direct electromyograph neurophysiological measures (Vecchio et al. 2020). Although the groups did not differ on neurophysiological variables before treatment, spasticity and pain scores improved in the MS patients during treatment by clinical scales and neurophysiological scales. These results suggest that a subset of MS patients is likely to find relief of MS symptoms from nabiximols.

Despite these positive findings for nabiximols in MS patients, there have been no human clinical trial studies of CBD alone on these symptoms. Since preclinical evidence suggests that $CB_1$ receptor agonists reduce, and $CB_1$ receptor antagonists enhance, hind-limb spasticity in mouse models of MS, it is likely that THC is the active agent in Sativex responsible for the reduction of these symptoms (Wilkinson et al. 2003). However, a recent preclinical finding with mice suggests that daily treatment with a 1 percent pure CBD propylene glycol cream may exert neuroprotective effects against autoimmune encephalitis (AEA), a model of MS (Giacoppo et al. 2015). The treatment began at the time of symptomatic disease onset, and mice were observed daily for signs of AEA (paralysis of hind limbs) and weight loss. One month after AEA induction, spinal cord and spleen tissue were examined. The results indicated that the CBD treatment reduced paralysis of hind limbs and lymphocytic demyelination in spinal cord tissues and reduced release of proinflammatory cytokines and markers of oxidative injury, suggesting that CBD alone may be useful in management of MS and its associated symptoms. (See also chapter 7 for further discussion of MS pain.)

**Alzheimer's Disease (AD)**

AD is characterized by a decline in cognitive and intellectual function that interferes with daily living. The brains of patients with AD show accumulation of amyloid-β protein in extracellular senile plaques in various brain regions, particularly the hippocampus, cerebral prefrontal cortex, and amygdala. A second marker of AD is the presence of intracellular neurofibrillary tangles consisting of hyperphosphorylated tau protein. These disorders of brain function are believed to be produced by neuroinflammation and oxidative stress (Ahmed et al. 2015; Watt and Karl 2017).

Numerous in vitro and in vivo studies have shown that cannabinoid agonists (such as THC) reduce amyloid β neurotoxicity by $CB_1$ and $CB_2$ receptor mechanisms of action. After subchronic administration for three weeks, CBD has also been shown to prevent amyloid-β-induced microglial activation both in vitro and in vivo (Martin-Moreno et al. 2011). CBD inhibits in vitro and in vivo amyloid-β peptide toxicity and inhibits the hyperphosphorylation of tau protein in vitro in a neuronal cell line (Aso et al. 2015). Finally, CBD also reduced memory impairments produced by amyloid-β in mouse models (Aymerich et al. 2018; Martin-Moreno et al. 2011). Most recently, chronic CBD (5 mg/kg, i.p., daily for three weeks) was found to reverse object recognition deficits in AD transgenic

(APPxPS1) female mice (Coles et al. 2020). However, there have been no human clinical trials with CBD as a treatment for AD patients.

**Parkinson's Disease (PD)**
PD is characterized by progressive degeneration of dopamine (DA) neurons of the substantia nigra that can be triggered by genetic risk in combination with unknown environmental factors (Fernández-Ruiz et al. 2014). It is a motor disease affecting 1 to 2 percent of the population aged sixty-five years or older. $CB_1$ and $CB_2$ receptor agonists have shown promise in various in vitro and in vivo models. The preclinical evidence for the effectiveness of CBD to restore DA levels in rat models of PD is very limited, with one study showing that CBD restored DA levels but another showing it was ineffective (Aymerich et al. 2018). There have been no published studies of CBD as a treatment for PD in human clinical trials, but there has been a recently published open-label, dose-escalation study of the safety and tolerability of CBD (Epidiolex) in PD (Leehey et al. 2020). CBD was titrated from 5 to 20 to 25 mg/kg/day and maintained for ten to fifteen days. All thirteen PD patients reported mild adverse effects including diarrhea, somnolence, and fatigue. Elevated liver enzymes occurred in five participants on the highest dose. Ten of the patients who completed the study showed improvement in movement disorder and improved sleep. Randomized controlled trials are needed to investigate various forms of cannabis in PD.

**Huntington's Disease (HD)**
HD is an inherited progressive neurodegenerative disease. A mutation in the huntingtin gene (IT15) results in damage to neurons in the striatum, resulting in movement and cognitive disorders. Early clinical trials with CBD aimed at treating motor dysfunction were without success (Fernández-Ruiz et al. 2014; Consroe et al. 1991) despite in vitro evidence that CBD may be neuroprotective in this disease by acting as an antioxidant, decreasing oxidative injury

(Fernández-Ruiz et al. 2014). A small-scale double-blind, randomized, cross-over, placebo-controlled clinical trial (NCT01502046) with twenty-six patients with HD did not show a benefit of nabiximols (Sativex: 2.7 mg THC/2.5 mg CBD, oromucosal spray, for twelve weeks) in motor, cognitive, behavioral, or functional outcomes compared with placebo in HD patients. (Lopez-Sendon Moreno et al. 2016). Clinical trials to date do not support the use of CBD in treating HD in humans.

## Conclusion: CBD and Neurodegenerative Disorders

There are no properly controlled human clinical trials for CBD in any neurodegenerative disorder and very little preclinical evidence for its effectiveness in treating these disorders. The most promising preclinical data are for the treatment of AD, but again those data are limited and confined to in vitro and in vivo preclinical data. Sativex (THC:CBD) is an approved treatment for tremors and neuropathic pain in MS patients in Canada and in Europe; however, the preclinical evidence suggests that the therapeutically beneficial substance for this disease is THC, not CBD.

## CBD and Skin Disease

CBD in dermatology (Is it hope or hype?) has been recently reviewed (Nickles and Lio 2020). The current literature indicates that CBD may be beneficial in skin disease, particularly in the treatment of acne, chronic pruritus (itchy skin), and atopic dermatitis, but the studies tend to be small and lacking rigorous design. There is a need for high-quality randomized, controlled trials to fully evaluate the efficacy and safety of CBD compounds before they can be promoted to treat dermatological diseases.

CBD products can be purchased online, over the counter, and at cannabis dispensaries. It has been marketed to consumers as being anti-inflammatory and analgesic. Others claim it to be a cure for acne, eczema, psoriasis, and pruritus. Still others claim that it has cosmetic value to reduce skin aging, increase skin moisturizing, and even increase hair-moisturizing. In fact, these products have been endorsed by celebrities, with the global skin care market estimated to be worth between $135 and $155 billion by 2021(Jhawar et al. 2019). However, many of these claims are unsubstantiated. To date, there have been very few published human clinical trials on the efficacy of CBD to treat skin disorders, but currently four relevant human trials are listed on the NIH website, www.clinicaltrials.gov: NCT04045314: CBD and 1 percent hemp oil on the hydration and erythema (redness) of skin; NCT04045119: Facial cream containing CBD and hemp oil on skin hydration and acne prone skin; NCT03573518: Evaluation of BTX 1503 (CBD formulation) in patients with moderate to severe acne vulgaris; NCT03824405: Study of the safety, tolerability and efficacy of BTX 1204 (CBD formulation) in patients with moderate atopic dermatitis (eczema). Results have not yet been posted for any of these trials.

**Regulatory Concerns of Commercially Available CBD Products**
In the United States at the state level, CBD products are considered legal if sold in states that allow for medical and recreational marijuana or in all states if extracted from hemp. However, at the federal level, the issue of legality of CBD is more complicated. The Drug Enforcement Administration (DEA) and the Controlled Substances Act regulate psychoactive compounds with abuse potential. Marijuana and all of its cannabinoid components have traditionally been considered to be Schedule I substances, with no currently accepted medical treatment use and deemed to have a high potential for abuse. However, the FDA has recently clarified that it does not consider CBD to be equivalent to marijuana by approving Epidiolex for treatment of childhood epilepsy. Additionally, the Department of Agriculture and the Farm Bill state that CBD extracted from hemp

is considered legal. Therefore, many proponents argue that CBD, especially from hemp, is legal and should be widely available.

This conflict between the state and federal regulations has created confusion for clinicians, patients, and consumers. The widespread availability of products containing CBD, when paired with this confusion about product legality, has added to the controversy. Skin care companies are continuing to offer products to consumers as if they are legal; however, these products have not been tested for safety or efficacy by the FDA and therefore lack regulatory approval. Products differ in terms of purity, strength, and source of the CBD (Jhawar et al. 2019).

**Transdermal Delivery of CBD**

Oral CBD has poor bioavailability (13–19 percent; Mechoulam, Parker, and Gallily 2002), suggesting that transdermal delivery may be preferable, especially when used to treat diseases of the skin. However, CBD is a highly lipophilic molecule and tends to accumulate within the upper skin layer, the stratum corneum, and poorly permeates to deeper layers (Lodzki et al. 2003). Transport into deeper layers requires efficient permeation enhancement. One technique for permeation enhancement is the use of ethosomes, phospholipid nanovescicles used for dermal and transdermal delivery of molecules (Touitou et al. 2000). A transdermal ethosomal CBD (6 mg) delivery system was applied by a patch transdermally for seventy-two hours to the abdomen of nude mice, which resulted in significant accumulation of CBD within the skin and underlying muscle. Steady state was reached in twenty-four hours with approximate dose accumulating in plasma being 45 mg/kg at twelve hours and 87 mg/kg after the entire seventy-two hours. This ethosomal CBD prevented inflammation and edema induced by subplantar (hindpaw) injection of carrageenan in these mice. Therefore, when administered in ethosomes, CBD permeates the skin such that it can be used as a transdermal anti-inflammatory treatment.

**Psoriasis and Eczema**
Keratinocytes are the predominant cell type of the epidermis (outer layers of skin) and originate in the basal layer (fourth layer of the epidermis). These cells produce keratin and are responsible for the epidermal water barrier by making and producing lipids. Keratinocytes differentiate as they travel through the outer layers of the skin to the skin surface. Their main purpose is to preserve against microbial, viral, fungal, and parasitic invasion to protect against UV radiation and minimize heat, solute, and water loss. When keratinocytes strongly increase their proliferation, psoriasis—a common inflammatory skin disease resulting in the manifestation of unsightly lesions within the epidermis—can develop. Psoriasis is also accompanied by increased expression of pro-inflammatory mediators in the skin. Since CBD is known to have anti-inflammatory properties and has been reported to have an inhibitory effect on rapidly proliferating tumorigenic cell lines, it was tested for its ability to inhibit the proliferation of human keratinocyte cells (Wilkinson and Williamson 2007). Cells were bathed in CBD or vehicle for seventy-two hours. CBD was found to inhibit the proliferation of hyperproliferating human keratinocyte cells in a concentration-dependent manner with average $IC_{50}$ values of 2 µM. Therefore, CBD reduces proliferation of human keratinocytes in vitro (Wilkinson and Williamson 2007), suggesting that it might be a treatment for psoriasis.

The effect of CBD on the expression of skin differentiation genes was also investigated in human keratinocyte cells (Pucci et al. 2013). CBD significantly reduced the expression of human keratinocyte cell genes by increasing DNA methylation of the keratin 10 gene. These results suggest that CBD acts as a transcriptional repressor that can control cell proliferation and differentiation.

The effect of CBD on the redox balance and phospholipid metabolism in UVA/UVB-irradiated keratinocytes isolated from skin of psoriatic patients or healthy volunteers was recently examined (Jarocka-Karpowicz et al. 2020). CBD partially reduces oxidative stress in keratinocytes of healthy individuals, while showing a tendency to increase the oxidative stress in keratinocytes of patients with psoriasis, especially following UV radiation. These changes induced by CBD treatment may be the cellular mechanism for CBD's effects in psoriasis. Using a system biology approach, Casares et al. (2020) characterized the effects of CBD on primary human keratinocytes at the molecular level. A functional analysis revealed CBD-regulated pathways involved in keratinocyte differentiation, skin development, and epidermal cell differentiation among other processes. CBD was found to induce the expression of several antioxidant pathways in keratinocytes and to regulate the physiological and pathophysiological outcomes of oxidant exposure on these skin cells. Collectively, these in vitro findings suggest that CBD has the potential to be tested in preclinical and eventually human clinical trials as a novel therapeutic for some skin diseases, including psoriasis. However, the single preclinical in vivo study on the effect of CBD on keratinocytes in mice demonstrated that CBD *increased* (rather than decreased) the level of some pro-proliferation genes (keratin 16 and 17) indicated in hyperproliferation in psoriasis (Casares et al. 2020). Therefore, the authors suggest that the use of CBD in psoriasis should be taken with caution due to its pro-proliferative effects in vivo but suggest that their findings indicate that CBD may be beneficial in other skin conditions such as eczema. Clearly, more preclinical work is warranted before CBD should be taken into human clinical trials for the treatment of psoriasis or other skin disorders.

### Acne Vulgaris

The most common human skin disease is acne vulgaris. Current acne treatments target multiple pathogenetic steps of acne, including sebum overproduction, unwanted sebocyte proliferation, and inflammation. There has been recent research into the cutaneous endocannabinoid system seeking novel therapeutic tools. The skin endocannabinoid system regulates cutaneous cell growth and differentiation and regulates sebum production, and it exerts anti-inflammatory effects (Olah et al. 2014). Indeed, anandamide is produced in human sebaceous glands with a function in the production of lipid (sebum) sebocytes.

CBD was evaluated for its effect on sebocyte production using human (SZ95) sebaceous gland cells. CBD (1–10 µM) normalized excessive lipid synthesis in human sebocytes induced by a pro-acne agent (Olah et al. 2014). Besides the lipostatic action, another desired effect of an anti-acne agent would be to inhibit the unwanted growth of sebocytes. CBD decreased proliferation of human sebocytes induced by a pro-acne agent; however, it did not reduce the number of sebocytes at baseline. Thus, CBD reduced proliferation but not the viability of these cell lines. Of clinical importance is whether these in vitro findings can be translated to cells from human sebaceous glands in situ. Using a model that mimics the human sebaceous gland function, CBD completely prevented the lipogenic effect of anandamide and decreased basal lipogenesis as well. CBD also suppressed the expression of the proliferation marker, MK167, suggesting that it would be effective in protecting against the sebocyte proliferation of acne (Olah et al. 2014). Both the antilipostatic effect and the antiproliferation effect of CBD on sebocytes was found to be mediated by its action on transient receptor potential vanilloid 4 (TRPV4) ion channels, as previous reports have shown that CBD acts on this receptor (De Petrocellis et al. 2012).

These promising in vitro findings suggest that CBD may be a potential treatment for acne vulgaris. One should keep in mind that topical administration may be preferable as well as using appropriate vehicles already used in current standard acne management. Due to its high lipophilicity, CBD is expected to preferentially enter the skin via the transfollicular route and to accumulate in the sebaceous glands (Lodzki et al. 2003); however, there is currently no preclinical animal model to evaluate the effectiveness of CBD in vivo to treat acne. As well, there are no randomized, double-blind, placebo-controlled, human clinical trials published for the efficacy of CBD for this condition.

**Protection from UV Radiation**

Exposure to solar UV radiation is a causative factor in acute skin photodamage, chronic photoaging, and photocarcinogenesis. Ultraviolet (UVA and UVB) radiation commonly causes a redox imbalance with a shift toward oxidation as a result of increased reactive oxygen species (ROS) generation that, when overproduced, can lead to cell damage. One ongoing area of interest is the potential of CBD to reduce the impact of harmful UV radiation (Jastrzab, Gegotek, and Skrzydlewska 2019). CBD is a well-known anti-inflammatory and antioxidant compound. CBD was reported to enhance the activity of antioxidant enzymes in UV-irradiated keratinocytes. Recently CBD was shown to reverse the metabolism of skin keratinocytes produced by UVA and UVB irradiation in nude rats (Atalay et al. 2021). However, there is no human clinical trial evidence indicating that CBD is effective as a protectant against UV radiation.

**Melanoma**

Skin cancer, the most common form of cancer, consists of melanoma and nonmelanoma skin cancers. Malignant melanoma, a solid tumor of pigment-producing cells in the skin, is the most lethal form of skin cancer. New advances in research for treatment of this aggressive cancer include immunotherapy and therapies developed to target the mutated BRAF gene in melanoma. However, melanoma continues to be difficult to treat, with current treatments producing significant side effects. CBD has been investigated for its potentially beneficial effect in treatment of melanoma in mice in an experimental cancer model (Simmerman et al. 2019). Murine melanoma cells (B16F10) were injected subcutaneously in mice, and the size of the growing tumor was measured using a caliper over time. When tumor size reached 3 mm in the longest diameter, treatments were performed.

When the tumor size reached 12 mm, the mice were euthanized as an experimental end point. The treatments included placebo control, CBD (5 mg/kg, i.p.), and a common anticancer treatment, cisplatin. CBD increased the time for the tumor to reach 12 mm and reduced the melanoma tumor growth rate relative to placebo, but cisplatin was more effective than CBD. Cisplatin, unlike CBD, however, is known to produce considerable side effects that impact quality of life. These findings are promising, suggesting a need for human clinical trials to evaluate the potential of CBD as a combination treatment, possibly with immunotherapy or BRAF-targeted therapy for melanoma.

**CBD in Hair Growth**
Although a number of CBD-infused shampoos and conditioners are on the market, there is little evidence to suggest that they are taken up by the hair but instead are likely taken up by the skin of the scalp. This would require the shampoo or conditioner to remain on the scalp long enough to be absorbed. These products are touted as being able to strengthen hair, prevent hair breakage and frizz, and even to promote hair growth. To date only one in vitro study has evaluated CBD's effect on hair growth using human scalp hair follicles and keratinocytes from plucked hairs from dermatologically healthy individuals. The results suggest that CBD at low concentrations (less than 1 μM) has minor effects on the hair growth cycle, but higher concentrations (10 μM and over) actually decrease hair growth, with the latter effect potentially being attractive for curbing unwanted hair (Szabó et al. 2020). Therefore, the effect of CBD on hair growth, at least for dermatologically healthy individuals, may be intimately tied to how much CBD is in the shampoo or conditioner and how much is actually penetrating the scalp.

**Conclusion: CBD and Skin**
Despite the considerable claims on the Internet about the potential of CBD to improve skin quality and reduce diseases of the skin, most

of the scientific evidence is in the form of in vitro studies on cell lines. For most of these diseases, reliable and valid preclinical animal models have not been developed for experimental evaluation of the claims. Only now are human clinical trials beginning with four currently posted on www.clinicaltrials.gov; however, no results of these trials have been posted or published.

## Cancer

There is considerable in vitro and more limited preclinical in vivo evidence suggesting that CBD exhibits pro-apoptotic and antiproliferative actions in several tumor lines, including human breast cancer, prostate and colorectal carcinoma, gastric adenocarcinoma, and rat glioma and transformed thyroid cells (Ligresti, De Petrocellis, and Di Marzo 2016). CBD may also exert antimigratory, antiinvasive, antimetastatic and perhaps antiangiogenic properties (Kovalchuk and Kovalchuk 2020; Pisanti et al. 2017; Massi et al. 2013). The efficacy of CBD is linked to its unique ability to target multiple cellular pathways controlling tumorigenesis through the modulation of different intracellular signaling pathways depending on cancer type.

### CBD and Breast Cancer

CBD potently and selectively inhibited the growth of different breast tumor cell lines with an $IC_{50}$ of about 6 µM, without inhibiting the growth of noncancer cells (Ligresti et al. 2006). As well, CBD inhibited the growth of breast tumors derived from injection of human breast cancer cells in mice, as well as infiltration of lung metastases derived from these cells. The mechanism for these effects involved direct TRPV1 activation and/or $CB_2$ indirect activation, as well as induction of oxidative stress. A later study found that in addition to proliferation, CBD also interfered with cell progression, invasion, and metastasis, all important hallmarks in the control of cancer (McAllister et al. 2007). CBD has been shown to induce a concentration-dependent cell death of both estrogen receptor-positive and estrogen receptor-negative breast cancer cells, with little effect on non-tumorigenic mammary cells, resulting in selective apoptosis and autophagy (cell death) in cancer cells, by an increase in ROS production.

## CBD and Glioma

CBD also reduces gliomas, brain tumors of glial origin, considered one of the most devastating neoplasms, with a high proliferative rate, aggressive invasiveness, and insensitivity to radiation and chemotherapy. CBD was first shown to reduce glioma cell proliferation in vitro in a mouse cell line (Jacobsson et al. 2000) and then in a human cell line (Massi et al. 2004). CBD was also shown to reduce glioma tumor growth in vivo in immune-deficient mice (Massi et al. 2004). The antitumor effect of CBD involved the induction of oxidative stress through increased early production of ROS; the CBD antiproliferative effect was reversed by the antioxidant tocopherol. Importantly, CBD did not induce ROS production in normal cells. Subsequently, it was noted that a combined treatment with CBD and THC in vitro reduced human glioma cell viability, enhancing autophagy and apoptosis (Torres et al. 2011), as well as reducing the growth of glioma tumors in mice.

## CBD and Leukemia and Lymphoma

CBD has been shown to induce apoptosis in a human leukemia cell line without affecting noncancer cells (Gallily et al. 2003). It has also been shown to induce apoptosis both in vitro and in vivo in murine and human leukemia cell lines (McKallip et al. 2006) by an increase in reactive oxygen species production and activation of $CB_2$ receptors. Indeed, human leukemias and lymphomas express significantly higher levels of $CB_2$ receptors compared with other tumor cell lines, indicating that these tumors, which arise from the immune system, may be especially sensitive to the $CB_2$-mediated effects of CBD (Massi et al. 2013).

### CBD and Lung Cancer
Lung cancer is aggressive and poorly responds to conventional therapy. There is in vitro evidence that CBD may reduce the invasive properties of human lung carcinoma cells, which express both $CB_1$ and $CB_2$ receptors, as well as TRPV1 receptors (Ramer and Hinz 2017). Additionally, in vivo studies in mice revealed an inhibition of lung cancer cell metastases following CBD treatment, suggesting that CBD may be a therapeutic option for this type of cancer.

### CBD and Thyroid Cancer
Thyroid cancer is the most common endocrine malignancy. CBD exerted antiproliferative effects (Ligresti et al. 2006) and induced apoptosis (Lee et al. 2008) in vitro in rat thyroid cancer cells by inducing oxidative stress and activation of ROS.

### CBD and Colon Cancer
Both in vitro and in vivo preclinical research suggest that CBD may be a possible treatment of colon cancer. Using a mouse model of colon cancer (Izzo et al. 2009), CBD was shown to reduce the development of polyps (growths on the inner surface of the colon from which nearly all colon cancers begin) and tumors.

### CBD and Angiogenesis
Angiogenesis is the formation of new blood vessels from preexisting ones, and the formation of cancer depends on this process, representing another promising therapeutic target to stop cancers from growing. Collectively, cannabinoids have been shown to act as anti-angiogenic factors by direct modulation of endothelial cells. CBD also affected endothelial cell differentiation in vitro and produced anti-angiogenic effects in vivo (Massi et al. 2013).

## CBD as an Extracellular Vesicle Inhibitor

A recent report evaluated the potential of CBD to act as an extracellular vesicle (EV) inhibitor (Kosgodage et al. 2018). EVs are key mediators in cellular communication for the transfer of proteins and genetic material. Cancers use EV release for drug-efflux, pro-oncogenic signaling, invasion, and immunosuppression. EV inhibitors have been shown to increase sensitivity of cancer cells to chemotherapy. CBD acts as an EV inhibitor. The potential of CBD to modify the EV profile in glioblastoma cells (the most common and aggressive form of primary malignant brain tumor in adults) in the presence and absence of temozolomide (chemotherapy treatment) was evaluated (Kosgodage et al. 2019). Compared to controls, CBD-treated cells released EVs containing lower levels of pro-oncogenic markers and increased levels of anti-oncogenic markers. These effects of CBD were greater than with the chemotherapy drug temozolomide alone. These results suggest that CBD may act as an adjunct to enhance treatment efficiency in glioblastoma cancer by modulation of EVs.

## CBD and Human Cancer Trials

Based on preclinical studies that demonstrated a synergistic effect of THC/CBD combinations and the chemotherapeutic drug temozolomide on glioma cells (Torres et al. 2011), a phase 2 randomized controlled, small-scale trial with twenty-one patients diagnosed with recurrent glioblastoma investigated the effects of Sativex/nabiximols versus placebo in combination with temozolomide. The preliminary results showed increased one-year survival (83 percent versus 53 percent). A recent clinical study evaluated the effect of pharmaceutical-grade CBD on a range of cancer patients over a four-year period. Clinical responses were seen in 92 percent of the 119 cases with solid tumors, including a reduction in circulating tumor cells in many cases and a reduction in tumor size in other cases, with no adverse effects observed. These are encouraging preliminary findings, but no double-blind, placebo-controlled study has been conducted to demonstrate the efficacy of CBD in any human cancers (Kenyon, Liu, and Dalgleish 2018). There are currently only two clinical trials listed on www.clinicaltrials.gov for treatment of cancer with CBD. One is a randomized, double-blind, placebo-controlled multicenter study to assess the efficacy of CBD combined with standard-of-care treatment in participants with multiple myeloma, glioblastoma multiforme, and gastrointestinal malignancies (Trial NCT-3607643) at Southwest Cancer Center in Orlando, Florida (PI Sara Katta, funded by Leaf Vertical), which is not yet recruiting patients. The other, described as "Pure CBD as single agent for solid tumor" (NCT02255292), by Yakir Rotenberg at Hadassah Medical Center in Israel was posted in 2014, but this trial is not yet recruiting participants.

**Conclusion: CBD and Cancer**

Several in vitro and preclinical in vivo studies using a variety of cancer cell lines and mouse models suggest that CBD may control both cancer growth and cancer spread. The anticancer effects of this compound seem to be selective for cancer cells, at least in vitro, since it does not affect normal cell lines. The efficacy of CBD is linked to its ability to target multiple cellular pathways and different intracellular signals depending on cancer type. The most common effect of CBD is the increase in ROS production to facilitate cancer cell death that seems to be a trigger for its beneficial action in all cancer cell types (Singer et al. 2015). The role of cannabinoid receptors mediating CBD's anticancer effects is more controversial. Some in vitro evidence suggests that specific antagonists of cannabinoid receptors may contribute to lung cancer, leukemia, and colon cancer but not glioma and breast cancer. These in vitro and preclinical in vivo results suggest that CBD may be worthy of clinical consideration for cancer therapy (Massi et al. 2013); however, no human-randomized, properly controlled trials of CBD in comparison to known treatments for any form of cancer have been conducted. We expect that over the next five years, several new clinical trials for the efficacy of CBD in treating cancer in humans will be conducted.

## General Conclusion

The currently available in vitro and preclinical in vivo data suggest that CBD may be neuroprotective and tissue protective, but important human clinical data are lacking for each of these indications. CBD may be an important treatment option in the neuroprotection against stroke, heart and liver injury, and bowel disorders. Preclinical animal studies also suggest that CBD may be a useful treatment in autoimmune diseases such as diabetes, hepatitis, and myocarditis. As well, it may play a role in protection against the neurodegenerative diseases AD, MS, and HD. However, human-randomized, placebo-controlled, clinical trials for the efficacy of CBD to treat any of these disorders have not been conducted. Nevertheless, promising human clinical trial data suggest that CBD may decrease the autoimmune reaction of graft versus host disease in transplant patients.

The evidence for the use of CBD to treat skin disorders is limited almost exclusively to in vitro evidence in cells. There is considerable promising in vitro evidence that CBD may reduce the proliferation of keratinocytes in eczema and psoriasis and sebocytes in acne vulgaris; however, currently no preclinical animal model has evaluated the effectiveness of CBD in vivo to treat these skin disorders. As well, there are no randomized, double-blind, placebo-controlled human clinical trials published for the efficacy of CBD to treat skin diseases.

In the past few decades, significant studies have investigated the safety, efficacy, and clinical utility of cannabis and cannabinoids in preclinical and clinical cancer models. These have also been evaluated for the management of cancer-associated symptoms such as nausea and vomiting (reviewed in the next chapter). In vitro and very limited in vivo preclinical research suggests that CBD exhibits pro-apoptotic and antiproliferative actions in several tumor lines, including human breast cancer, prostate and colorectal carcinoma, gastric adenocarcinoma, and rat glioma, and transformed thyroid cells. CBD may also exert antimigratory, anti-invasive, antimetastatic, and perhaps anti-angiogenic properties. Most clinically available cancer drugs induce systemic toxicity in patients, which may limit the therapeutic utility of many of these drugs. However, CBD has a relatively lower level of toxicity to such patients. The question will be whether human clinical trials support some of the in vitro and preclinical in vivo evidence reviewed in this chapter about the efficacy of CBD in cancer chemotherapy patients.

# 6
# Nausea, Vomiting, and Appetite

Nausea and vomiting are common symptoms that may occur as a side effect of medications or treatments (as in the case of cancer chemotherapy or radiation therapy) or due to a disorder (as in Crohn's disease). Nausea is typically described as a subjective feeling of an unsettled stomach accompanied by the sensation that vomiting is imminent. Although it can accompany nausea, emesis, commonly known as vomiting, is a separate process involving the expulsion of stomach contents through the mouth.

Appetite loss and malnutrition are often also associated with serious illnesses such as cancer. Cachexia—a disorder that causes extreme weight loss, muscle wasting, and body fat loss—affects people who are in the late stages of serious diseases. Despite available treatments and symptoms management, cancer patients may still experience nausea, vomiting, appetite loss, and/or cachexia, greatly impacting their quality of life. Given these unmet treatment needs, the exploration of complementary or alternative treatments, such as cannabinoids, is needed.

## Chemotherapy- or Radiation-Induced Nausea and Vomiting

Nausea and vomiting are adverse side effects of chemotherapy or radiation that oncology patients often experience despite the use of current antiemetic treatments (Navari and Aapro 2016). These symptoms impact treatment outcome and patient quality of life, highlighting the need to manage nausea and vomiting.

There are three types of chemotherapy-induced nausea and vomiting. Nausea and vomiting that occur within twenty-four hours of the initial anticancer agent administration are classified as *acute*, while that occurring twenty-four hours (often peaking two or three days) after administration of the anticancer agent is classified as *delayed*. If the acute and delayed phases are not properly managed, *anticipatory* nausea and vomiting can result, whereby arrival at the cancer clinic triggers an episode of nausea or vomiting, or both. This is due to a learned association between the cues of the clinic (such as sights and smells) and the later experience of nausea and vomiting from the anticancer agent administered.

While there is a good understanding of the neurobiology of vomiting, nausea is less clear. Toxic chemotherapy agents (or radiation) stimulate enterochromaffin cells in the gut. This causes the release of serotonin, which then binds to serotonin 3 ($5\text{-}HT_3$) receptors of intestinal vagal afferent nerves, triggering acute vomiting. In contrast, the delayed phase of vomiting occurs due to the release of substance P from neurons in the central and peripheral nervous systems. Substance P activates neurokinin-1 (NK1) receptors, resulting in delayed vomiting.

The guidelines for the management of nausea and vomiting typically include a three-drug regimen consisting of a $5\text{-}HT_3$ receptor antagonist (ondansetron, dolasetron, and granisetron), an NK1 receptor antagonist (aprepitant, fosaprepitant, and rolapitant) and a corticosteroid (dexamethasone), with the addition of the anxiolytic agent olanzepine for highly toxic chemotherapy agents. Although these medications are good at managing vomiting, they are minimally effective in reducing acute and delayed nausea (Hickok et al. 2003; Poli-Bigelli et al. 2003), and none of these medications are effective in reducing anticipatory nausea. Nausea (acute, delayed, and anticipatory) continues to be a problematic symptom for oncology patients, underscoring the need for the evaluation of alternative treatments, such as cannabis and cannabinoid compounds. Maida and Daeninck (2016) provide an excellent overview of cannabinoid use to manage symptoms in oncology patients.

In this chapter, we focus on evidence for whole plant cannabis (in which CBD is present), nabiximols (which contains THC and CBD in a 1:1 ratio), and a few studies that have evaluated CBD alone, for its antinausea and antivomiting role, with evidence at the level of in vitro, in vivo, and human trials. Next, we discuss the effects of CBD on appetite and weight since cancer patients, as well as those with many chronic illnesses such as AIDS, experience loss of appetite, body weight loss, and malnutrition. Finally, this chapter briefly discusses the condition of cannabinoid hyperemesis, a syndrome characterized by severe episodes of nausea and vomiting in those using cannabis for prolonged periods, particularly with high doses of THC.

## CBD and 5-HT Involvement in Nausea and Vomiting: In Vitro Effects

The $5\text{-HT}_3$ receptor antagonists typically prescribed to chemotherapy patients exert their action by decreasing 5-HT transmission, suggesting that other drugs that have the capacity to reduce 5-HT may also have antinausea or antivomiting potential. Indeed, in vitro studies have investigated CBD's action at 5-HT receptors.

Using frog cells expressing mouse and human $5\text{-HT}_{3A}$ receptors, researchers have investigated CBD's effects on 5-HT (Yang et al. 2010). CBD was inhibitory and reversed 5-HT-induced currents in a concentration-dependent manner ($IC_{50} = 0.6$ μM). Although the potency of 5-HT was not altered, CBD did decrease its efficacy, suggesting allosteric modulation of $5\text{-HT}_3$ receptors by CBD. This allosteric binding of CBD to the $5\text{-HT}_3$ receptor would lead to less 5-HT signaling (as would also occur with a $5\text{-HT}_3$ receptor antagonist), and this decreased 5-HT signaling may be a mechanism for CBD's antinausea and antivomiting properties.

Another receptor that reduces 5-HT signaling is the $5\text{-HT}_{1A}$ autoreceptor, located on the soma and dendrites of 5-HT neurons, where they exert negative feedback on subsequent firing (Blier et al. 1998). That is, $5\text{-HT}_{1A}$ autoreceptor activation leads to an inhibition of further 5-HT release and therefore may reduce nausea or vomiting, or both. Early studies determined that CBD (at the rather high concentration of 16 µM) activated human $5\text{-HT}_{1A}$ receptors transfected into Chinese hamster ovary cells (Russo et al. 2005). More recently, the binding of CBD at lower concentrations to $5\text{-HT}_{1A}$ receptors in rat brainstem membranes in vitro was examined (Rock et al. 2012). CBD (100 nM, but not at 10 nM or 1 µM) strengthened the ability of the $5\text{-HT}_{1A}$ receptor agonist, 8-OH-DPAT, to stimulate binding. In addition, CBD did not act as a direct agonist at the $5\text{-HT}_{1A}$ receptor, and it was not able to alter 8-OH-DPAT's rate of dissociation from binding sites in these membranes (Rock et al. 2012). Together, these findings suggest that CBD may have an indirect action by enhancing the action of endogenous 5-HT at the $5\text{-HT}_{1A}$ receptor.

The acidic precursor of CBD, CBDA also enhanced the binding of 8-OH-DPAT to $5\text{-HT}_{1A}$ receptors but did so over a much wider range of concentrations (0.1 to 100 nM) in comparison to CBD (Bolognini et al. 2013). The methyl ester version of CBDA, HU-580 (0.01 to 10 nM) potently enhanced the binding of 8-OH-DPAT to human $5\text{-HT}_{1A}$ receptors (Pertwee et al. 2018). This suggests that CBD (as well as CBDA and HU-580) may suppress nausea or vomiting, or both, through activation of $5\text{-HT}_{1A}$ receptors, ultimately enhancing the endogenous 5-HT's ability to bind to these autoreceptors and shut down further release of illness-inducing 5-HT.

## CBD and 5-HT Involvement in Nausea and Vomiting: In Vivo Effects

### Antiemetic Effects

Common laboratory species used to screen potential antiemetic compounds are shrews, ferrets, and cats. These animals vomit in response to the administration of toxins such as lithium chloride (LiCl) and the chemotherapeutic agent cisplatin, as well as in response to rotation (motion-induced vomiting). In squirrel monkeys, coadministration of CBD (3 mg/kg, i.m.) reduced high-dose (1 mg/kg, i.m.) THC-induced emesis (Withey et al. 2021).

The antiemetic effects of CBD in toxin- and motion-induced vomiting have been investigated in the house musk shrew, *Suncus murinus*. In the shrew, CBD produces a biphasic effect, such that low doses (5, 10 mg/kg, i.p., s.c.) reduce (Rock et al. 2011; 2012), while high doses (25, 40 mg/kg, i.p.) potentiate (but do not themselves produce) toxin-induced vomiting (Kwiatkowska et al. 2004; Parker, Kwiatkowska et al. 2004). CBD's antiemetic effects are mediated by the 5-HT$_{1A}$ receptor (Rock et al. 2012), but not by the CB$_1$ receptor (Kwiatkowska et al. 2004). Repeated CBD (5 mg/kg, s.c., seven days) treatment maintained its antiemetic efficacy in shrews, suggesting that they did not become tolerant to CBD's antiemetic effect (Rock, Sullivan, Collins et al. 2020).

In comparison to CBD, CBDA (0.1 and 0.5 mg/kg, i.p.) potently reduced toxin-induced vomiting (Bolognini et al. 2013), and recent unpublished data suggest that unlike CBD, CBDA (20 mg/kg, i.p.) does not potentiate toxin-induced vomiting. When ineffective antiemetic doses of CBD (2.5 mg/kg, i.p.) or CBDA (0.05 mg/kg, i.p.) were combined with a subthreshold antiemetic dose of THC (1 mg/kg, i.p.), synergistic effects were seen (Rock and Parker 2015).

These findings indicate that CBD and CBDA are effective toxin-induced, antiemetic compounds in shrews, likely mediated by action at 5-HT$_{1A}$ receptors. CBD may have a narrow therapeutic window.

This might suggest that lower doses of CBD or CBDA may reduce toxin-induced vomiting. Whether the methyl version of CBDA is more potent in reducing emesis has not yet been evaluated.

CBD (0.5, 1, 2, 5, 10, 20, and 40 mg/kg, i.p.) is ineffective in reducing motion-induced vomiting in shrews (Cluny et al. 2008), while CBDA (0.1 and 0.5 mg/kg, i.p.) is effective (Bolognini et al. 2013). This inconsistency in CBD reducing toxin-induced, but not motion-induced vomiting, may be due to differing neuronal pathways involved in the induction of emesis and also the differing intensity of sickness. Motion-induced vomiting may be more severe and may require a more potent antiemetic, such as CBDA, to overcome it.

**Antinausea Effects**

To screen antinausea compounds, a reliable preclinical model of nausea is necessary. Rodents cannot vomit, but they still receive the same gut signals in response to illness-inducing agents (such as LiCl), presumably experiencing the nausea that often precedes vomiting (Billig, Yates, and Rinaman 2001), making them an ideal rodent nausea model. Parker and colleagues have provided considerable evidence that conditioned gaping—large openings of the mouth and jaw, with the lower incisors exposed—in rats, first described by Grill and Norgren (1978), is a selective measure of nausea (Parker 2014). Conditioned gaping is produced only by manipulations that produce vomiting in other species, and treatments that reduce nausea and vomiting in other species consistently prevent conditioned gaping. These conditioned gaping reactions are not only displayed to nausea-paired flavors (a rodent model of acute nausea) but also to nausea-paired contextual cues (a model of anticipatory nausea experienced by patients receiving chemotherapy treatment on their return to the clinic; Limebeer et al. 2008). We first present evidence for the preclinical antinausea effect of CBD (and CBDA and HU-580) in acute nausea, followed by evidence in anticipatory nausea.

In a rodent model of acute nausea, CBD (5 mg/kg, s.c. or i.p.) reduced conditioned gaping, suggesting an acute antinausea effect (Rock et al. 2012; Parker, Mechoulam, and Schlievert 2002; Parker and Mechoulam 2003). Recently the dose response was expanded, demonstrating a wider therapeutic window for CBD's antinausea effect from 0.5 to 5 mg/kg (i.p.), with no antinausea effect at 0.1, 20, and 40 mg/kg (i.p.) (Rock, Sullivan, Pravato et al. 2020). Importantly, these higher doses of CBD did not promote conditioned gaping but were simply ineffective. CBD maintained its antinausea effect when given repeatedly (5 mg/kg, s.c., daily for seven days or over four weekly treatments; Rock, Sullivan, Collins et al. 2020). The combination of ineffective doses of CBD (0.5 mg/kg, s.c., or 0.1 mg/kg, i.p.) with ineffective doses of the $5\text{-HT}_{1A}$ agonist 8-OH-DPAT (Rock et al. 2012) or THC (0.1 mg/kg, i.p.; Rock, Sullivan, Pravato et al. 2020) synergistically reduced conditioned gaping. This suggests that CBD can be administered at low doses along with other low-dose antinausea compounds to interfere with conditioned gaping.

CBD's effects in the acute conditioned gaping model seem to be mediated by its action at $5\text{-HT}_{1A}$ receptors (Rock et al. 2012). When injected into the dorsal raphe nucleus (a brain region responsible for forebrain 5-HT and the location of somatodendritic $5\text{-HT}_{1A}$ autoreceptors), CBD (10 μg) reduced conditioned gaping and this effect was blocked by administration of a $5\text{-HT}_{1A}$ receptor antagonist (Rock et al. 2012). The interoceptive insular cortex (IIC), a critical forebrain region mediating nausea, receives projections from the dorsal raphe nucleus. When rats are administered illness-inducing LiCl, selective elevation of 5-HT is detected in the IIC, and CBD (5 mg/kg, i.p.) reduced this 5-HT elevation (Limebeer et al. 2018). This suggests that the antinausea properties of CBD seem to be driven by its action at $5\text{-HT}_{1A}$ somatodendritic autoreceptors in the dorsal raphe nucleus to reduce the firing of 5-HT neurons with projections to the IIC. This inhibition prevents the elevation of 5-HT in the IIC from LiCl. In short, 5-HT in the IIC promotes the sensation of nausea, which may be prevented by CBD.

CBDA (0.5, 1, 5, 10, 100 μg/kg, i.p.) potently reduced conditioned gaping in this acute nausea model (Rock and Parker 2013) through a 5-$HT_{1A}$ receptor-mediated effect (Rock, Limebeer, and Parker 2015; Bolognini et al. 2013). When given orally, CBDA (2.5, 10, 20 μg/kg, but not 0.5, 1 μg/kg) also reduced conditioned gaping (Rock et al. 2016). Like CBD, repeated treatment with CBDA (1 μg/kg, s.c.) maintained its ability to reduce conditioned gaping (Rock, Sullivan, Collins et al. 2020). Combinations of an ineffective dose of CBDA (0.1 μg/kg, i.p.) with an ineffective dose of metoclopramide (a dopamine receptor antagonist used as an antiemetic that pre-dates the 5-$HT_3$ receptor antagonists; Rock and Parker 2013), the 5-$HT_3$ receptor antagonist ondansetron (1.0 μg/kg; Rock and Parker 2013), THCA (Rock, Sullivan, Pravato et al. 2020) or THC (Rock, Limebeer, and Parker 2015; Rock et al. 2016) produced an enhanced antinausea effect. Finally, HU-580 (0.1, 1 μg/kg) has been shown to be more effective than CBDA in reducing acute nausea, also through a 5-$HT_{1A}$ receptor mechanism (Pertwee et al. 2018), and maintains its efficacy with repeated dosing (Rock, Sullivan, Collins et al. 2020). This suggests that these cannabinoids may be potential antinausea treatments to investigate for the management of chronic conditions, without the development of tolerance.

In a rodent model of anticipatory nausea, CBD (1, 5, but not 10 mg/kg, i.p.) reduced (Rock et al. 2008) and CBDA (1, 10, 100 μg/kg, i.p.) more potently reduced contextually elicited conditioned gaping (Rock, Limebeer, and Parker 2015; Bolognini et al. 2013; Rock et al. 2014). At an extremely low dose of 0.01 μg/kg (i.p.), HU-580 was more effective than CBDA in reducing contextually elicited conditioned gaping (Pertwee et al. 2018), and these antinausea effects of CBDA and HU-580 were mediated by the 5-$HT_{1A}$ receptor (Bolognini et al. 2013; Pertwee et al. 2018). The combination of ineffective doses of CBDA (0.1, 1, 10 μg/kg, i.p.) with ineffective doses of THC

(1, 10 mg/kg, i.p.) or THCA (5 µg/kg, i.p.) enhanced the suppression of anticipatory nausea (Rock, Limebeer, and Parker 2015; Rock et al. 2014). Oral CBDA (10 µg/kg) also reduced anticipatory nausea (Rock et al. 2016), and when ineffective oral doses of CBDA (0.1 or 1 µg/kg) and THC (0.1 or 1 mg/kg; Rock et al. 2016) were combined, there was an enhanced suppression of conditioned gaping.

Together, the acute and anticipatory antinausea effects of CBD (with greater efficacy for CBDA and more so for HU-580) suggest that these compounds may be antinausea treatment candidates, worthy of further research. Their effectiveness in reducing anticipatory nausea is particularly important because no specific treatments for this disorder are currently available. Preclinical evidence suggests that low doses of CBD (and CBDA and HU-580) can be combined with other antiemetic compounds to interact synergistically to reduce nausea more effectively, potently, and perhaps selectively.

## Nausea and Vomiting in Humans

The anecdotal accounts from cancer patients who testified that smoking cannabis helped them to manage their nausea and vomiting prompted oncologists to begin to examine its efficacy. Reports have also indicated that cannabis use prior to a chemotherapy appointment decreased anticipatory nausea (Wilkie, Sakr, and Rizack 2016). As a result, a number of studies examining both dronabinol and nabilone in the 1970s and 1980s in chemotherapy patients were conducted and have been evaluated in numerous meta-analyses and systematic reviews (Whiting et al. 2015; Tramer et al. 2001; Machado Rocha et al. 2008). These reviews suggest that dronabinol and nabilone were superior to placebo and as effective as or superior to the antiemetics available at the time. In 1985, nabilone (Cesamet), an orally active, synthetic analogue of THC, was licensed for management of chemotherapy-induced nausea and vomiting and appetite promotion.

A report by the National Academy of Sciences (2017) indicated that there was conclusive evidence for the effectiveness of oral cannabinoids for the treatment of chemotherapy-induced nausea and vomiting. Reports from palliative care physicians who discuss cannabis use with their cancer patients suggest that approximately 50 percent of them intend to use cannabis to manage their nausea (Panozzo et al. 2020) Indeed, a survey of Canadian cancer patients indicated that 34 percent of those who reported cannabis use were doing so to manage their nausea (Martell et al. 2018). Despite patient perceptions of efficacy and their seeking of information, there is still a lack of data to support using plant cannabis to reduce nausea and vomiting.

**Nabiximols (THC+ CBD) as a Treatment for Nausea and Vomiting**
A pilot phase 2 randomized, double-blind, placebo-controlled trial has evaluated nabiximols (mean daily dose 13 mg THC+12 mg CBD for four days, beginning within two hours postchemotherapy administration) as an adjunct treatment in sixteen adult patients undergoing chemotherapy (Duran et al. 2010). Indeed, 71 percent of patients reported no delayed vomiting (versus 22 percent in placebo), and 57 percent reported no delayed nausea (versus 22 percent in placebo), but there were no differences in the acute period. No significant changes were seen in either group in terms of blood pressure, weight, temperature, hematology, or blood chemistry. There were no differences in quality-of-life measurements. This study suggests antiemetic and antinausea benefits with the addition of nabiximols to chemotherapy patients' treatment regimens but should be interpreted cautiously due to the small sample size.

A multicenter, randomized, double-blind, placebo-controlled, phase II trial evaluated oral THC+CBD (THC 2.5 mg+CBD 2.5 mg, three times daily, from days −1 to 5, and one cycle of matching placebo in a crossover design) extract for the prevention of chemotherapy-induced nausea and vomiting (Grimison et al. 2020). In comparison to placebo, the addition of the cannabis extract to the antiemetic regimen significantly reduced vomiting and nausea and improved quality-of-life scores (Grimison et al. 2020). Adverse events such as sedation, dizziness, and disorientation were reported, although 85 percent of participants preferred the extract over placebo. Based on these results, this group is continuing recruitment for the phase 3 parallel component of the trial.

**CBD as a Treatment for Nausea and Vomiting**
Only one case report has examined the effect of CBD treatment (100–450 mg/day, oral capsule, for two years) on chemotherapy-induced nausea (Dall'Stella et al. 2019). During their anticancer treatment period, two male patients with gliomas reported little nausea or fatigue and were even able to engage in sports. Importantly these patients also showed no evidence of disease progression for at least two years and did not exhibit any significant blood count or plasma biochemistry abnormalities. Although these results seem promising, future randomized, placebo-controlled trials with a larger number of patients are needed to properly evaluate the effects of CBD in managing nausea and vomiting in cancer patients.

**Nausea and Vomiting as Side Effects of CBD treatment**
Vomiting can be reported as an adverse effect associated with CBD treatment. A study of 607 treatment-resistant epilepsy patients showed that those treated with the lowest dose of CBD (2–10 mg/kg/day) as an adjunct treatment reported no vomiting, but increased incidences of vomiting occurred at higher doses (above 40 mg/kg/day; Szaflarski et al. 2018). These findings suggest that it may be higher doses of CBD that produce vomiting, but this is difficult to ascertain because patients also continued taking their other antiepileptic medications during these trials. It is possible that CBD (perhaps at higher doses) may produce vomiting, or it may be interacting with an adjunct treatment to produce this effect. In addition, it

cannot be verified that the CBD product is indeed CBD and does not contain contaminants. A meta-analysis of CBD treatment-associated adverse events reported in randomized clinical trials (twelve trials contributed data from 803 participants) did not find a significant increased odds of experiencing nausea or vomiting in those treated with CBD, relative to placebo (Chesney et al. 2020).

## Modulation of Appetite and Body Weight by CBD

*Cachexia* is a common term for the wasting symptoms that may appear in chronic illnesses, such as HIV/AIDS and cancer. This can be compounded by a loss of appetite and malnutrition in oncology patients. The data supporting cannabis and cannabinoid use in appetite stimulation is mixed, with some studies showing a benefit in terms of appetite stimulation and weight gain, while others show no effect or even appetite reductions.

### In Vitro Studies

One study has examined the effect of CBD-regulating levels of factors that control feeding. Rat hypothalamic Hypo-E22 cells and isolated hypothalamus tissues were exposed to CBD (1, 10, 100, 1000 nM; di Giacomo et al. 2020), modulating these food intake factors in a manner that would suggest an appetite loss effect (di Giacomo et al. 2020). Further research is needed here.

### Preclinical Studies

CBD treatment in rodents has been shown to have differential effects on intake of standard rodent show or sweet solutions such as sucrose or saccharin. Further studies are needed to parse out these conflicting findings.

The earliest report of CBD's effects on appetite by Sofia and Knobloch (1976) suggested a CBD-induced reduction of appetite. They showed that an acute high dose of CBD (50 mg/kg, i.p.) reduced chow and water consumption in rats (Sofia and Knobloch 1976). Similarly, acute administration of a moderate dose of CBD (4.4 mg/kg, oral) reduced chow consumption in rats, with lower doses of CBD (0.044, 0.44 mg/kg, oral) having no effect in nondeprived rats (Farrimond, Whalley, and Williams 2012). In addition, repeated administration of CBD (2.5 and 5 mg/kg, i.p. for fourteen days) reduced body weight gain dose-dependently in nondeprived rats (Ignatowska-Jankowska, Jankowski, and Swiergiel 2011). Finally, twice-daily CBD (20 mg/kg, i.p.) treatment during the adolescent period (25–45 days) resulted in reduced weight gain in female mice (Kaplan et al. 2021). Overall, these studies suggest that CBD may reduce appetite in rats (and perhaps female mice) with access to standard rodent chow.

Conversely, no effect on food intake because of CBD administration has been reported. CBD (1, 10, 20, i.p.) did not change food intake in deprived or nondeprived rats (Scopinho et al. 2011). Similarly, CBD (10 mg/kg, i.p.) had no effect on food consumption or water intake in nondeprived mice (Riedel et al. 2009), and in deprived mice, CBD (3, 10, 30, 100 mg/kg, i.p.) had no effect on food intake (Wiley et al. 2005). In contrast, repeated high-dose CBD (30 mg/kg, but not 3, 10 mg/kg, i.p., daily for fourteen days) treatment increased weight gain in diabetic rats (Chaves et al. 2020). Together, these mixed results do not reveal a clear role for CBD in food intake, but they do generally suggest an appetite-reduction effect (or no effect on appetite) to standard rodent chow. In a disease state such as diabetes, a high repeated dose of CBD (but not lower doses) did promote weight gain.

When rodents are additionally given access to a sweet solution, acute CBD (50 mg/kg, i.p.) produced a preference for sweet sucrose, but did not stimulate total appetite, as rats compensated by consuming less chow overall (Sofia and Knobloch 1976). In contrast, a low dose of CBD (2.5 mg/kg, i.p., daily for fourteen days) did not alter rats' liking responses to sweet saccharin (O'Brien, Limebeer et al. 2013), but it is unknown whether a high dose of CBD would alter such responses. Although the potent acidic version of CBD, CBDA (0.05–5 mg/kg oral), did not modify normal feeding behavior in rats (Brierley et al. 2016), CBDA (0.01 mg/kg, i.p. and oral) did increase liking responses to sweet saccharin (Solinas et al. 2013; Rock et al. 2016), suggesting that perhaps a high dose of CBD may also increase liking of sweet solutions. These findings suggest that rather than stimulating appetite in general, CBDA (and perhaps higher doses of CBD) may specifically stimulate liking of sweet, highly palatable foods. Whether CBD promotes the liking of sweet foods in humans, as it seems to in rats, remains to be determined.

A recent study has examined oral CBD oil in healthy dogs and assessed body weight and food intake. In this randomized, blinded, placebo-controlled trial, healthy dogs received a CBD (1, 2, 4, or 12 mg/kg; from cannabis extract) or placebo oil formulation orally once daily for twenty-eight days (Vaughn, Paulionis, and Kulpa 2021). Over the four-week treatment period, there was a mild mean reduction in body weight across all groups (including the placebo group; –0.7 percent), with a greater body weight reduction in the 1 (–4.4 percent), 2 (–5.7 percent), 4 (–4.7 percent) mg/kg CBD group (but not the 12 mg/kg CBD group; –2.3 percent), that was not accounted for by changes in food consumption or daily activity. It is unclear whether the body weight loss was due to the CBD or the medium-chain triglycerides (MCT) oil that the compound was mixed in, although Vaughn, Kulpa, and Paulionis (2020) have demonstrated that MCT but not sunflower oil can result in reduced food intake in dogs. Although this randomized, placebo-controlled, blinded, parallel study showed a reduction in food intake in healthy dogs treated with CBD oil and MCT oil, body weights remained stable throughout the study. These findings suggest that in dogs, CBD dose and the medium that it is mixed in are important factors mediating body weight loss and appetite. It is also important to note that this body weight loss is occurring in healthy dogs orally administered CBD (and MCT), which could be exacerbated in dogs with conditions that already are affecting their appetite and/or body weight.

**Human Studies**

Data from cannabis users generally suggest a promotion of appetite and weight gain in a variety of conditions, but a report by the National Academy of Sciences (2017) indicated insufficient evidence to support or refute the conclusion that cannabis or cannabinoids are an effective treatment for cancer-associated cachexia. Improved appetite or weight gain has been demonstrated in Crohn's disease patients who smoked cannabis cigarettes (115 mg of THC and less than 0.5 percent CBD, twice daily for eight weeks; Naftali et al. 2013); in advanced cancer patients treated with oral cannabis capsules (THC 9.5 mg and CBD 0.5 mg, once or twice daily, for six months; Bar-Sela et al. 2019); and in AIDS patients smoking cannabis (Haney et al. 2005, 2007).

It has been suggested, however, that it is likely that the THC in whole plant cannabis may be stimulating appetite in these studies rather than the CBD, as nabilone and dronabinol have been shown to improve appetite and/or promote weight gain in conditions such as anorexia nervosa, lung cancer, HIV/AIDS and Alzheimer's disease (Turcott et al. 2018; Andries et al. 2015, 2014; DeJesus et al. 2007, Wilson, Philpot, and Morley 2007, Volicer et al. 1997). Dronabinol is approved by the US Food and Drug Administration for the treatment of HIV wasting. Indeed, self-reported appetite stimulation increased with higher THC strains (Brunt et al. 2014). When viewing pictures of food (versus neutral objects such as car keys), intoxicated smokers of CBD-rich strains were less interested in pictures of food in comparison to smokers of THC-rich strains (Morgan, Freeman et al. 2010). Increased appetite or weight gain improvements with the treatment of nabiximols have also been reported in patients with chronic pain (Ueberall, Essner, and Mueller-Schwefe 2019), cannabis withdrawal (Allsop et al. 2014), and Huntington's disease (Saft et al. 2018). A small ($n=21$) two-arm prospective open-label pilot trial of escalating doses of CBD oil (100 mg/mL, oral, dose range 50 to 600 mg/day) or THC oil (10 mg/mL, oral, dose range 2.5 to 30 mg/day) in palliative care patients reported improved appetite overall but did not differentiate whether this improvement was more prevalent in the THC or CBD group (Good et al. 2020). A survey of outpatient palliative care patients in Florida indicated that 50 percent of those using THC products and 29 percent of those using CBD products reported improved appetite (Highet et al. 2020).

A few studies have suggested that CBD treatment may actually reduce appetite in patients. A recent meta-analysis of the adverse events associated with CBD treatment in randomized clinical trials (twelve trials contributed data from 803 participants) suggests that the odds of experiencing a decrease in appetite are three and a half times more likely for those treated with CBD versus placebo (Chesney et al. 2020). This adverse effect was most often reported in Dravet syndrome and Lennox-Gastaut syndrome patients treated with CBD and may be dose dependent such that higher doses may decrease appetite (Devinsky, Patel, Cross, Villanueva, Wirrell, Privitera, Greenwood, Roberts, Checketts, VanLandingham, and Zuberi 2018). It is difficult to ascertain if this is an effect due to CBD treatment itself or an interaction between CBD and the other antiepileptic medications that these patients are also taking during these studies. This potential adverse event should be a consideration for physicians when determining CBD doses.

## Cannabinoid Hyperemesis Syndrome and Potential Relief by CBD

Cannabinoid hyperemesis syndrome is associated with prolonged cannabis use, particularly with high THC content, and is characterized by cyclical episodes of nausea and vomiting, accompanied by abdominal pain (Allen et al. 2004). For an excellent review see

DeVuono and Parker 2020. Typical antiemetic drugs are ineffective in these patients (Richards 2017), but the application of capsaicin cream (containing the spicy chemical in chili peppers) on the abdomen (Dezieck et al. 2017) and taking hot showers can reduce symptoms (Wallace, Martin, and Park 2007). Dysregulation of TRPV1 receptors may be one of the factors contributing to this syndrome, as these receptors are activated by both of these treatments (Rudd et al. 2015; Richards 2017). Because CBD also activates TRPV1 (De Petrocellis et al. 2012), it could be evaluated as a treatment for cannabinoid hyperemesis syndrome, or, instead, smoking cannabis strains with higher CBD content (and therefore lower THC content) may be preventive. Indeed, CBD treatment was effective in a preclinical rodent model of THC-induced nausea (DeVuono et al. 2020). A recent study of twenty-eight cannabinoid hyperemesis syndrome patients revealed mutations in genes affecting neurotransmitters, the endocannabinoid system, and the cytochrome P450 complex associated with the metabolism of cannabinoids (Russo et al. 2021). Although quite preliminary, these potential genetic mutations may serve as a diagnostic signal to identify those who may be more susceptible to developing this syndrome. Further research is needed to understand the consequences of high dose cannabis use and the role that CBD may play in this syndrome.

## Conclusion

Vomiting and especially nausea are symptoms that are not well managed in oncology patients. Appetite loss is also an important factor in the well-being of these patients. Classic antiemetic compounds exert their action by decreasing 5-HT transmission, and in vitro and in vivo studies suggest that CBD (as well as CBDA and HU-580) might also exert their antinausea and antiemetic effects in this manner by acting at the $5\text{-HT}_{1A}$ receptor.

In animal models, CBD seems to have a narrow therapeutic window for its antiemetic effects in shrews, with low to moderate doses reducing vomiting but higher doses increasing toxin-induced vomiting. This might suggest the use of lower doses of CBD to treat toxin-induced vomiting. CBD reduces acute nausea (at a wide range of doses and does not promote nausea at higher doses) and anticipatory nausea in rats. CBDA and HU-580 potently reduce acute and anticipatory nausea in rats. Preclinical evidence also suggests that low doses of CBD (and CBDA and HU-580) can be combined with other antiemetic compounds and synergistically reduce acute and anticipatory nausea. This may be an important therapeutic advantage, as lower doses of these drugs are less likely to result in adverse side effects. Tolerance to the antinausea and antiemetic effects of repeated CBD treatment does not seem to occur in this animal model.

Human clinical trials suggest that THC and its analogs reduce chemotherapy-induced nausea and vomiting, but only one case report has assessed the effect of CBD on two patients who received chemotherapy and radiation, albeit with favorable antinausea effects. These results must be interpreted with caution as randomized, placebo-controlled trials are still needed.

The effects of CBD on appetite and body weight are mixed. In vitro findings suggest that CBD modulates food intake factors in a direction that suggests a loss of appetite and preclinical results suggest that CBD may reduce appetite (or have no effect) in rats consuming standard rat chow. But CBD may promote liking when rats are given access to sweet solutions, suggesting a specific liking of sweet solutions rather than a general effect on appetite. As well, CBD's effects on food intake and body weight in healthy dogs are difficult to interpret, as the oil itself that these compounds are administered with can affect these parameters. Further research needs to clarify these findings.

In humans, cannabis treatment is associated with appetite stimulation and weight gain, but this effect is likely mediated by THC rather than CBD. In fact, CBD treatment can produce the adverse event of decreasing appetite, an effect that has been reported in Dravet syndrome and Lennox-Gastaut syndrome patients and may be due to high-dose CBD treatment. Finally, cannabinoid hyperemesis syndrome, characterized by cyclical episodes of nausea and vomiting in high-THC-content chronic cannabis users, is treated by hot showers and capsaicin cream—treatments that activate TRPV1 receptors. Because CBD also activates these receptors, further research could investigate if CBD may be a potential treatment for this distressing syndrome. Alternatively, consumption of cannabis strains with higher CBD content may prevent its occurrence. Future research should evaluate the role that CBD may play in modifying chemotherapy-induced nausea and vomiting, appetite, and cannabinoid hyperemesis syndrome.

Some clinical trials are underway to explore the effects of cannabinoids on the symptoms associated with chemotherapy. A registered clinical trial (NCT03984214) is currently assessing the efficacy of dronabinol in improving chemotherapy-induced symptoms (such as appetite promotion and nausea and vomiting) in pancreatic cancer patients. In addition, two registered clinical trials (NCT03664141 and NCT04001010) are assessing the effects of cannabis oil (starting dose of 1.2 mg CBD+1.2 mg THC) on appetite in patients with end-stage kidney disease or the inhalation of synthetic THC+CBD on weight gain and quality-of-life measures in patients with advanced cancer. A randomized, double-blind, placebo-controlled study (NCT03948074) by Pippa Hawley at the British Columbia Cancer Agency will investigate the use of cannabis oil extracts (high THC/low CBD, low THC/high CBD, equal amounts THC/CBD) for the management of cancer symptoms, including nausea, pain, anxiety, and sleep disturbance. As well, a study by Jens Rikardt Andersen at the University of Copenhagen (NCT04585841) will monitor the effects of CBD on lean body mass during chemotherapy, as well as the management of nausea and vomiting. These important clinical trials will provide more information about the effects of CBD in managing symptoms associated with cancer treatment.

# 7
# Pain and Inflammation

Although our capacity to feel pain may seem highly undesirable at times, it is in fact protective and essential for our survival. Pain and inflammation are the body's response to injury, helping us avoid harm and further injury. Pain can be experienced due to activation of pro-inflammatory agents such as cytokines—including interleukins (IL), tumor necrosis factor alpha (TNF-α), and interferon gamma (IFNγ)—which mount an immune response and promote inflammation. Excessive, chronic production of these inflammatory cytokines contributes to the development of inflammatory diseases. (See chapter 5 for CBD's effects in the immune system, including inflammation and inflammatory diseases.)

Pain and inflammation can be classified in a number of different ways—for example, as acute or chronic. Acute pain and inflammation may be caused by surgery, trauma (such as a broken bone or cut), or muscle strain. It can last up to weeks or months, but it typically ends once its underlying cause has been treated or healed. In contrast, chronic pain and inflammation are persistent and last for at least three months, extending beyond the time of normal tissue healing.

Pain and inflammation can also be classified as nociceptive or neuropathic. Nociceptive pain and inflammation are experienced as a result of tissue injury or damage (for example, spraining your ankle or closing the car door on your fingers), while neuropathic pain and inflammation are experienced as a result of nerve injury or disease. People experience neuropathic pain after having shingles or following sciatic nerve or spinal cord injury. Neuropathic pain patients often experience symptoms such as allodynia (pain due to a stimulus that does not normally provoke pain) and hyperalgesia (excessive pain from a stimulus that would usually provoke milder pain).

Three major drug classes are typically used to treat pain and inflammation: acetaminophen (Tylenol), nonsteroidal anti-inflammatory drugs such as aspirin and ibuprofen, and opioids. Nonsteroidal anti-inflammatory drugs combat pain by reducing inflammation, while opioids combat pain by blocking pain signals from traveling to the brain. Each of these drugs is associated with differing side effects and with varying degrees of efficacy. Although opioids offer the strongest analgesic (pain-relieving) effect, they have significant addictive potential, making the development of opioid alternatives essential. Also of interest is the identification of drugs that can boost the analgesic effects of opioids (known as opioid-sparing effects), allowing patients to achieve similar analgesic properties with the use of a lower opioid dose.

There is significant interest in determining whether medical cannabis, or particular cannabinoids such as CBD, could replace opiates as analgesics or act as opioid-sparing compounds. It has been suggested that patients should be cautious when taking CBD in combination with other drugs because it inhibits several cytochrome P450 isoenzymes (Balachandran, Elsohly, and Hill 2021). For the management of chronic pain, this may be especially important because typical analgesics are metabolized by the P450 isoenzymes CYP2D6 and CYP3A4. CBD inhibits CYP2D6 and CYP3A4 and can therefore increase bodily levels of drugs that are normally metabolized by these isoenzymes. In fact, in vitro studies suggest that CBD inhibits the breakdown of the opiate heroin (Qian, Gilliland,

and Markowitz 2020), which could result in higher levels of heroin in the body when taken with CBD. This suggests a biological basis for the possibility of CBD producing opioid-sparing effects by augmenting levels of opioids in the body; however, this finding was not supported in a double-blind, placebo-controlled cross-over study (Manini et al. 2015) that examined the safety and pharmacokinetics of CBD (400 or 800 mg, oral capsule) coadministered with intravenous fentanyl, in seventeen healthy participants. CBD was well tolerated, but it did not produce any significant pharmacokinetic changes with fentanyl coadministration. No serious adverse events such as respiratory depression or cardiovascular compromise in any subject were reported. A case report of a thirteen-year-old cancer and chronic pain patient does suggest that CBD treatment may have augmented serum methadone levels. A patient using methadone with a CBD product presented to her physician with sleepiness and fatigue, accompanied by a serum methadone level of 271 ng/mL (Madden, Tanco, and Bruera 2020). Upon discontinuing CBD, the serum level dropped to 125 ng/mL, and sleepiness and fatigue diminished. The potential for CBD-opiate interactions is clinically important and warrants further study. Here we first review the preclinical studies investigating CBD's effects in pain, leading to a discussion of the few clinical trials that have begun to determine the potential pain-relieving effects of CBD.

## CBD and Pain: Preclinical Studies

Animal models of pain are used to assess the effects of potential therapeutic compounds, indexed by behavioral outcomes. These models do have limitations in their application to pain patients. For example, in contrast to the variability in pain patient background, the rodent strains used are genetically very similar, often using only males, which does not reflect the patient population.

Other limitations include the reliance on young, healthy animals and brief dosing and testing timelines that would reflect earlier, rather than late disease stages.

**CBD and Acute Pain Models**

Acute pain is typically assessed in rodents by applying stimuli such as heat (to the tail or paw) or pressure (to the paw). The latency to withdraw the paw or tail is measured in response to heat (in the hot plate test or tail flick test). The force necessary to produce a paw withdrawal is measured in response to applying gradually increasing pressures (Randall-Selitto paw pressure test or Von Frey filaments). If a compound increases this latency (it takes longer for the animal to detect the painful stimulus) or increases the required force (it takes more force for the animal to detect the painful stimulus), this compound is thought to have an analgesic effect.

Systemic injection or oral dosing of CBD in mice and rats does not have an analgesic effect in the heat and paw pressure tests of acute pain (Sofia, Vassar, and Knobloch 1975; Karniol and Carlini 1973; Varvel et al. 2006; Britch et al. 2017; Chesher et al. 1973; Silva et al. 2017); however, CBD has been shown to potentiate the analgesic effect of THC (Varvel et al. 2006; Britch et al. 2017) or morphine (Neelakantan et al. 2015) in some of these models. This potentiation of THC-induced analgesia by CBD is accompanied by an increase in elevated blood THC levels, likely due to CBD's ability to inhibit THC metabolism (Varvel et al. 2006; Britch et al. 2017). In contrast, repeated treatment with CBD (10 mg/kg, i.p., twice daily for four days) actually decreased THC's antinociceptive effect in the paw pressure test and tail flick test, with higher serum THC levels in the CBD and THC-treated rats (Greene et al. 2018). It is possible that the sustained elevation of blood THC levels due to repeated THC and CBD treatment leads to a dysregulation of the endocannabinoid system, ultimately reducing analgesia.

One study has demonstrated an antinociceptive effect of topical CBD cream (5 percent) in rats using the hot plate test of acute pain (Yimam et al. 2021). Here, rats treated with CBD cream showed a 38.4 percent increase in paw withdrawal latency compared to vehicle, indicating a pain-relieving effect. In contrast, treatment with 5 percent ibuprofen showed a 22.4 percent increase. Finally, when infused directly into the ventrolateral periaqueductal gray (a brain region involved in the modulation of nociception), CBD (1.5, 3, but not 6 nM) increased tail-flick latency, an analgesic effect (Maione et al. 2011). This effect was blocked by a $CB_1$ receptor antagonist, suggesting an involvement of the $CB_1$ receptor (Maione et al. 2011). This involvement of the $CB_1$ receptor is not direct activation by CBD but is likely due to CBD's known ability to inhibit FAAH (De Petrocellis 2012), thereby elevating anandamide (and other fatty acid amides), which activates the $CB_1$ receptor; however, recent findings suggest that CBD may not inhibit FAAH in human cells (Criscuolo et al. 2020). CBD's analgesic effect was also blocked by the 5-$HT_{1A}$ receptor antagonist WAY100635, suggesting involvement of the 5-$HT_{1A}$ receptor as well (Maione et al. 2011). In contrast, when administered directly into the cerebrospinal fluid in the cerebral ventricles by intracerebroventricular injection (i.c.v. infusion), in order to bypass the blood-brain barrier and distribute CBD throughout the brain, CBD (3 nmol, i.c.v.) did not have analgesic effects in the tail flick test, but when coadministered with morphine (6 nmol, i.c.v.), CBD potentiated its analgesic effect (Rodriguez-Munoz et al. 2018). Taken together, these results suggest that systemic injection of CBD is ineffective alone in acute pain models, but when combined with THC or morphine, it may potentiate their analgesic effects, although this beneficial effect may be diminished after repeated treatment. When administered topically or directly into the ventrolateral periaqueductal gray (a brain region involved in the modulation of nociception), CBD may be analgesic, suggesting that sufficient doses of

CBD to exert acute analgesic effects may not be reaching the brain regions crucial for pain, when administered systemically.

Two studies have assessed the effect of analogs of CBD in acute pain models in animals. Analogs of the (+)-enantiomer of CBD reduced analgesia in the hot plate test, without producing the typical psychoactive effects induced by $CB_1$ receptor agonists (Fride et al. 2004). As well, a fluorinated CBD analog (HUF-101) potently reduced pain in the hot plate test (Silva et al. 2017). Both a $CB_1$ receptor antagonist and a $CB_2$ receptor antagonist blocked HUF-101's analgesic effect in the hot plate test (Silva et al. 2017). In addition, HUF-101 did not produce the typical psychoactive effects induced by $CB_1$ agonists (Silva et al. 2017). These results suggest that HUF-101's effects may be complex in this model, with an involvement of both $CB_1$ and $CB_2$ receptors. Indeed, CBD's activity at these receptors is complicated; it can act as an indirect agonist of cannabinoid receptors by increasing endocannabinoid tone, as both a partial agonist at the $CB_2$ receptor (Tham et al. 2019), and as a negative allosteric modulator of the $CB_1$ and $CB_2$ receptors (Martinez-Pinilla et al. 2017). Further complicating CBD's interaction with the $CB_2$ receptor is that its partial agonist activity depends on the expression and density of the receptors in the area of interest, as well as the tonic activity of the endocannabinoid system (Mlost, Bryk, and Starowicz 2020). (For a detailed review of the various mechanisms of action mediating CBD's effects in pain models, see Mlost et al. 2020.) These findings suggest that CBD as well as these CBD analogs may be promising analgesic compounds to evaluate for acute pain, without intoxicating effects.

## CBD and Acetic Acid–, Formalin- or Phenyl Benzoquinone–Induced Pain and Acute Inflammation

Models that incorporate the application or injection of chemicals (such as formalin) leading to short-term or acute inflammation can also be used to assess pain. In these models, when the toxin is injected into the paw, the quantification of time spent elevating the paw or licking the paw is used as a measure of spontaneous pain behavior, with an analgesic compound reducing this behavior. When a toxin is injected intraperitoneally, writhing behavior is measured, with an analgesic compound reducing this behavior.

In general, systemic injection or oral dosing of CBD in mice and rats does not have an analgesic effect in the formalin- or acetic acid-induced test of acute pain and inflammation (Sofia, Vassar, and Knobloch 1975; Finn et al. 2004; Welburn et al. 1976; Booker et al. 2009; Yimam et al. 2021), but a handful of studies have demonstrated that CBD (30, 90 mg/kg, i.p., or 10 mg/kg, oral, or 5.6, 56 mg/kg, i.p.) reduced abdominal writhing to acetic acid (Silva et al. 2017; Foss et al. 2021) or phenyl benzoquinone (Formukong, Evans, and Evans 1988) in mice. It is possible that the level of pain induced in these latter studies was less severe and could be surmounted by CBD. In addition, analogs of the (+)-enantiomer of CBD, as well as fluorinated CBD (HUF-101) and the CBD analogue KLS-13019, also reduced pain in the formalin test (Fride et al. 2004) or the acetic acid test (Silva et al. 2017; Foss et al. 2021). Together, these studies suggest mixed results in terms of CBD's efficacy in models of short-term pain and acute inflammation. The ineffectiveness of CBD may be due to an insurmountable level of pain in some models.

## CBD and Neuropathic Pain Models

A common type of chronic pain is neuropathic pain associated with nerve injury. It is associated with exaggerated pain perception, which may be in the form of increased sensitivity to normal stimulation (hyperalgesia) or from stimuli that normally would not provoke pain (allodynia). These stimuli can be mechanical (an object touching the skin) or thermal (heat or cold applied). This increased pain sensitivity results in shorter paw withdrawal latencies or more paw lifts or licks in response to stimuli. Analgesic compounds reduce this enhanced pain sensitivity, resulting in longer paw withdrawal latencies (or fewer paw lifts or paw licks).

The first animal model of neuropathic pain was the chronic constriction injury of the sciatic nerve, in which ligatures are placed around the sciatic nerve to occlude blood flow, producing increased pain sensitivity of the hind paw. Animal models of neuropathic pain have since been developed that use the administration of toxic substances, such as chemotherapeutic agents. In addition, several models have been developed that mimic spinal cord injury pain, which also results in hyperalgesia and allodynia. There is also the recent development of animal models of disease-related peripheral neuropathic pain, including pain associated with diabetic neuropathy. We review evidence for the effect of CBD in each of these neuropathic pain models.

## Chronic or Partial Constriction Injury of the Sciatic Nerve Model of Neuropathic Pain

The majority of animal studies on neuropathic pain rely on traumatic injury to the sciatic nerve. In the chronic constriction model of neuropathic pain, acute treatment with CBD (30 mg/kg, s.c., 100 mg/kg, oral) reduced mechanical or cold pain sensitivity (Casey, Atwal, and Vaughan 2017; Mitchell et al. 2021). Similarly, with repeated administration following the chronic or partial constriction surgery, CBD (2.5–20 mg/kg, oral for seven days; 5 mg/kg, s.c. for seven days; 1 mg/15 ml of gelatin for three weeks) reduced thermal and mechanical pain sensitivity (De Gregorio et al. 2019; Comelli et al. 2008; Costa et al. 2007; Abraham et al. 2019). The effects of CBD were fully blocked by a selective TRPV1 antagonist (but not by $CB_1$ or $CB_2$ receptor antagonists; Comelli et al. 2008; Costa et al. 2007; De Gregorio et al. 2019). These findings suggest that activation of TRPV1 by CBD may lead to receptor desensitization and ultimately reduced neuropathic pain symptoms.

A modification of the partial sciatic lesion model, constriction by a cuff, leads to a shorter-lasting sciatic nerve injury model. Using this sciatic cuff model of neuropathic pain, CBD (approximately 0.4 mg/kg, oral, for fourteen days beginning 24 hours after surgery), or the methyl ester version of CBDA, HU-580 (1 µg/kg, i.p., for fourteen days beginning 24 hours after surgery), reduced thermal sensitivity in male rats, but were ineffective in females (Zhu et al. 2020; Linher-Melville et al. 2020). These results highlight the need for a comparison of CBD and CBDA's effects in male and female rodent models of sciatic nerve injury neuropathic pain, as well as other pain models. These initial findings with CBDA may suggest that higher doses of CBDA (and perhaps CBD) may be needed for pain relief in females.

**Chemotherapy-Induced Neuropathic Pain Model**

Chemotherapy agents can damage the peripheral nerves, causing pain and discomfort, often preventing patients from achieving effective chemotherapy doses during their treatment. This damage produces hyperalgesia and allodynia. When administered to animals, chemotherapy agents also produce enhanced pain sensitivity, allowing the study of prevention, as well as treatment of chemotherapy-induced neuropathic pain.

Paclitaxel, used to treat a variety of cancers, including ovarian and breast tumors, also results in enhanced pain sensitivity in paclitaxel-treated animals. CBD (1–20 mg/kg, i.p., for up to fourteen days) prevented the development of paclitaxel-induced cold or mechanical enhanced pain sensitivity in mice (Ward et al. 2011, 2014; King et al. 2017), and this effect was blocked by a 5-HT$_{1A}$ receptor antagonist, but not by a CB$_1$ or CB$_2$ receptor antagonist (Ward et al. 2014). Interestingly, when a very low ineffective dose of CBD (0.3 mg/kg, four days) was combined with a very low ineffective dose of THC (0.3 mg/kg, four days), this combination was synergistic and prevented paclitaxel-induced mechanical sensitivity (King et al. 2017). A structural analog of CBD, KLS-13019 (2.5 mg/kg, i.p.

or oral, four days), which shows superior bioavailability over CBD, prevents the development of paclitaxel-induced mechanical sensitivity (like CBD; Foss et al. 2021). Furthermore, although CBD treatment did not reverse already established peripheral neuropathy, KLS-13019 (2.5 mg/kg, i.p., oral, three days) did (Foss et al. 2021). Therefore, CBD may be effective (on its own or administered with a very low dose of THC) in preventing paclitaxel-induced pain sensitivity by its action at the 5-$HT_{1A}$ receptor. The structural analog of CBD, KLS-13019, may also prevent and reverse peripheral neuropathy in this animal model.

Platinum-based chemotherapy agents, including cisplatin and oxaliplatin, are commonly used to treat ovarian cancer and colorectal cancer. These agents produce mechanical sensitivity that can last more than ten years (Strumberg et al. 2002). CBD (10 mg/kg, i.p.) prevented oxaliplatin-induced mechanical sensitivity (King et al. 2017), and when a very low ineffective dose of THC (0.16 mg/kg, i.p.) was combined with a very low ineffective dose of CBD (0.16 mg/kg, i.p), this combination was synergistic and prevented oxaliplatin-induced mechanical sensitivity (King et al. 2017). In another study, CBD (2 mg/kg, i.p.) attenuated but did not prevent cisplatin-induced mechanical sensitivity (Harris et al. 2016).

In addition to producing neuropathic pain, cisplatin can also produce kidney damage, but CBD (10 mg/kg, i.p.) reduced inflammation and kidney injury when administered prior to cisplatin in a mouse model of cisplatin-induced nephropathy (Pan et al. 2009). Several markers of nephrotoxicity were also reduced in CBD-treated animals, demonstrating a beneficial effect of CBD against cisplatin-induced side effects.

Vincristine, typically used to treat leukemia and lymphomas, produces profound mechanical pain sensitivity. Treatment with CBD (1.25–10.0 mg/kg, i.p., seven days) was ineffective at preventing vincristine-induced mechanical sensitivity in mice (King et al. 2017). It is possible that CBD was ineffective due to a suboptimal dose of CBD used in the study or an insurmountable level of pain expressed in these mice. In summary, adjunct treatment with CBD may be effective in the prevention or attenuation of some chemotherapy-induced neuropathic pain in animal models. If CBD could prevent or reduce the pain associated with these treatments, cancer patients may be better able to tolerate optimal doses of these chemotherapeutic agents, promoting better cancer recovery prospects.

**Spinal Cord Injury Model of Neuropathic Pain**
Injury to the spinal cord can affect the transmission of sensory signals, causing the development of neuropathic pain, with the symptoms of allodynia, hyperalgesia, and spontaneous pain. Up to 45 percent of spinal cord injury patients experience neuropathic pain, often within three months of injury (Norrbrink Budh et al. 2003). This pain is often treated with antiepileptics, antidepressants, or opioids, but some of these treatments also have harmful side effects.

CBD (1.5 mg/kg, i.p., for ten weeks) attenuated the development of thermal sensitivity following spinal cord injury (Li et al. 2018). In addition, CBD treatment was associated with a decrease in pro-inflammatory cytokines, suggesting an anti-inflammatory effect (Li et al. 2018). CBD treatment was also associated with a decrease in other signaling molecules (such as T cells) involved in activating the immune response (Li et al. 2018). Because CBD treatment attenuated the development of thermal sensitivity and protected against inflammation and a pathological immune response, it may be a candidate adjunct treatment for evaluation in neuropathic pain associated with spinal cord injury.

**Diabetic Neuropathic Pain**
Diabetic neuropathic pain is characterized by tingling, burning, sharp, and shooting sensations in the hands and feet. The pain can be constant and accompanied by greater pain sensitivity. Pharmacological treatment such as anticonvulsants and antidepressants manage the symptoms, although unsatisfactorily for many patients.

Diabetes can be precipitated in mice by injection of streptozotocin, which kills insulin-secreting islet cells and initiates diabetes within one week. Diabetic mice develop tactile thermal sensitivity within three months of the streptozotocin injection.

Treatment with CBD (1, 2 mg/kg, intranasal or 20 mg/kg, i.p., for three months), beginning one week after the diabetes induction, limited the development of thermal and mechanical sensitivity in mice (Toth et al. 2010). In fact, CBD's analgesic effect was maintained for additional assessments over two months, even after cessation of CBD treatment. CBD treatment also prevented the development of molecular hallmarks of diabetes (Toth et al. 2010). But when given to mice that had already established this neuropathic pain, CBD was ineffective in reducing thermal hypersensitivity to tactile allodynia (Toth et al. 2010). These early findings suggested that CBD may be effective in preventing the development of diabetic neuropathic pain in mice if given at an early stage of diabetes.

Most recently, using the streptozotocin-induced diabetes model in rats, Jesus et al. (2019) showed that when administered twenty-eight days after the initiation of diabetes, CBD (0.3, 3 mg/kg, i.p.) reduced mechanical sensitivity, and this effect was blocked by the 5-HT$_{1A}$ receptor antagonist, but not by a CB$_1$ or CB$_2$ receptor antagonist. When administered fourteen days after the initiation of diabetes, repeated treatment with CBD (0.3 or 3 mg/kg, i.p. for fourteen days) induced a sustained attenuation of the mechanical sensitivity in the diabetic rats (Jesus et al. 2019). Analysis of the serotonin levels in the spinal cords of these rats revealed that CBD (0.3 mg/kg, i.p. for fourteen days) restored the reduced levels of spinal cord serotonin in these rats. Taken together, these results suggest that CBD may be effective in the treatment of diabetic neuropathic pain and, its effects may be mediated by the serotonergic system. Species differences may explain the differential effects of CBD. It appears that in mice, CBD may be effective in relieving neuropathic pain only when administered at an early stage in disease progression, while in rats, there may be a wider time line in which initiation of CBD treatment may still be beneficial.

## CBD and Postoperative Pain

Surgical incision of the rat foot causes a reliable and quantifiable mechanical sensitivity lasting for several days after surgery, producing a rodent model of postoperative pain. Using this model, systemic CBD (3 and 10 mg/kg, i.p.) or injection into the rostral anterior cingulate cortex (a brain region implicated in the processing of pain; 40 nmol), when administered twenty-four hours after surgery, reduced mechanical sensitivity (Genaro et al. 2017). These findings suggest that CBD may reduce postoperative pain in this animal model, and the rostral anterior cingulate cortex may be an important brain region involved in this type of pain.

## CBD and Inflammatory Pain Models

Animal models of inflammatory pain have used a number of different irritants injected into the skin, paw, muscle, joint, and visceral organs to produce acute inflammatory pain. These models produce inflammation at the site of injection, as well as greater pain sensitivity, and have been useful in screening potential analgesic/anti-inflammatory compounds such as CBD.

### Carrageenan-Induced Inflammatory Pain

Carrageenan, a commonly used inflammatory irritant, is typically injected into the paw to produce acute inflammation that converts to a chronic inflammation by two weeks. Carrageenan injection also leads to increased sensitivity to thermal and mechanical stimuli at the site of injection. This model has been used to understand inflammatory pain and likely mimics conditions associated with tissue injury like sprains and strains.

A single oral treatment of CBD (10, 20, 40 mg/kg, administered one hour after carrageenan) reduced mechanical sensitivity and also reduced the physical swelling (edema) of the paw in rats, measured two to six hours after carrageenan (Yimam et al. 2021). Further, the reduction in pain sensitivity and inflammation shown in rats treated with 40 mg/kg oral CBD was similar to that achieved by 100 mg/kg ibuprofen. When administered daily after the onset of inflammation by carrageenan, low doses of CBD (5, 7.5 mg/kg, oral for three days) reduced, and high doses of CBD (10, 20, 40 mg/kg, oral for three days; 10 mg/kg, oral once) abolished, thermal sensitivity in rats (Costa, Colleoni et al. 2004; Costa, Giagnoni et al. 2004). CBD (7.5–40 mg/kg, oral) also reduced the physical swelling of the paw after a single administration (Costa, Colleoni et al. 2004). The analgesic effect of CBD (10 mg/kg, oral) was blocked by a TRPV1 receptor-selective antagonist but not a $CB_1$ or $CB_2$ receptor-selective antagonist (Costa, Colleoni et al. 2004), suggesting an importance of TRPV1 in CBD's analgesic effect. CBD treatment also inhibited the overproduction of pro-inflammatory mediators in carrageenan-treated rats (Costa, Giagnoni et al. 2004). These results suggest that CBD has anti-inflammatory activity when administered after the onset of inflammation, indicating that it may have a significant therapeutic effect, which may be of benefit in inflammatory diseases.

The acidic precursor of CBD, CBDA also shows analgesic effects in preclinical models. Rock et al. (2018) showed that administration of CBD (10 mg/kg, oral) prior to carrageenan reduced thermal sensitivity in rats, as did CBDA, but at much lower doses. CBDA (10 µg/kg i.p. or 1, 100 µg/kg oral) prior to carrageenan (but not after carrageenan) reduced sensitivity and had anti-inflammatory effects (Rock, Limebeer, and Parker 2018). The analgesic effects of CBDA were blocked by a TRPV1 receptor antagonist. Interestingly, the combination of ineffective doses of CBDA and THC (100 µg/kg THC+0.1 µg/kg CBDA, oral) showed an enhanced effect in reducing thermal sensitivity and inflammation. Therefore, CBDA alone, as well as very low doses of combined CBDA and THC, have anti-inflammatory effects and reduce enhanced thermal pain sensitivity in this animal model of inflammation.

When administered prior to carrageenan, the fluorinated version of CBD, HUF-101 (3, 10, 30 mg/kg, i.p.). potently decreased mechanical sensitivity, as higher doses of CBD (30, 90 mg/kg, i.p.) were necessary to similarly reduce mechanical sensitivity in mice (Silva et al. 2017). The effects of CBD and HUF-101 were blocked by $CB_1$ and $CB_2$ receptor antagonists (Silva et al. 2017). Taken together, these findings suggest that HUF-101 and CBDA produced analgesic effects at lower doses than CBD. Furthermore, some of the analgesic effects of CBD (and HUF-101 and CBDA) seem to involve the activation of TRPV1 or the cannabinoid receptors. In some cases, CBD's effects were also blocked by $CB_1$ receptor antagonism, likely by CBD inhibiting FAAH and ultimately elevating levels of the endogenous cannabinoid anandamide, which does in fact readily bind to $CB_1$ receptors. Indeed, FAAH inhibition also reduces carrageenan-induced mechanical sensitivity via the $CB_1$ receptor in mice (Grim et al. 2014). CBD also acts as both a partial agonist (Tham et al. 2019) and a negative allosteric modulator of the $CB_2$ receptor (Martinez-Pinilla et al. 2017), which may also be contributing to its analgesic effects in these models.

**Complete Freund's Adjuvant-Induced Inflammatory Pain**

Complete Freund's adjuvant, typically injected into the paw or a specific joint, results in a more chronic inflammation than carrageenan and is accompanied by thermal and mechanical sensitivity. This model is routinely used for screening novel compounds for inflammatory pain, such as osteoarthritis pain, which occurs in the hands, knees, hips, and back when the cartilage is broken down in joints.

When Complete Freund's adjuvant was injected into the hind paw, CBD (20 mg/kg, oral treatment for seven days), treatment beginning on day 7 following the initiation of inflammation reduced thermal and mechanical sensitivity in rats (Costa et al. 2007). Also using this model, CBD (2.5, 10 mg/kg, i.p.) reduced mechanical sensitivity or decreased paw thickness (similarly in male and female rats) when administered acutely or repeatedly immediately following initiation of inflammation (as well as twice daily for three more days; Britch et al. 2020). These analgesic and anti-inflammatory effects of CBD were associated with an inhibition of pro-inflammatory mediators activated during inflammation (Britch et al. 2020).

Injection of Complete Freund's adjuvant directly into the knee joint causes joint swelling and thermal paw sensitivity. CBD (6.2, 62.3 mg/day, transdermal gel, beginning three days after induction of knee arthritis, for four days), reduced thermal paw sensitivity, and joint swelling (Hammell et al. 2016). CBD also reduced pro-inflammatory biomarkers, immune cell infiltration, and the pathological thickening of the synovium (the membrane lining the joints) that is seen with arthritis. Plasma analysis showed that transdermal administration of CBD was detectable in the blood. This suggests transdermal CBD may be a candidate for evaluation for treatment of arthritic joints.

Along with injection of Complete Freund's adjuvant, collagen can also be injected at the base of the tail to induce rheumatoid arthritis in mice. In collagen-induced arthritis, joint inflammation results and can be treated by blocking pro-inflammatory cytokines (such as TNF-$\alpha$), which are highly expressed in the arthritic joints of mice. CBD (5 mg/kg, i.p., daily for ten days once arthritis was established) reduced swelling and joint rigidity, with CBD treatment restoring the joints to normal in 34 percent of these mice (Malfait et al. 2000). Daily oral CBD (25 mg/kg) was also effective at slowing arthritis progression when administered for four weeks once arthritis was established (Malfait et al. 2000). The synovial membrane lining the joints is the most critical site of cytokine production in arthritis, and synovial cells from collagen-induced arthritic mice at day 10 are known to spontaneously produce large amounts of TNF-$\alpha$ when cultured. CBD (5, 10 mg/kg, i.p.) reduced TNF-α production by synovial cells and reduced a toxin-induced rise in serum TNF-α (Malfait et al. 2000). These outcomes suggest that CBD's beneficial effect in arthritis seems to be the result of anti-inflammatory action reducing TNF-α in the synovial membrane of the joint.

## Osteoarthritis Models

Osteoarthritis can be induced by injection of sodium monoiodoacetate into the knee joint, producing sensitivity in the hind paw of rats. Fourteen days after the initiation of arthritis, CBD (300 μg into the knee joint) treatment reduced mechanical sensitivity, and this effect was blocked by a TRPV1 antagonist (Philpott, O'Brien, and McDougall 2017). In addition, injection of CBD into the knee joint reduced inflammation (Philpott, O'Brien, and McDougall 2017). When administered prophylactically for three days, CBD (300 μg into the knee joint) also attenuated the development of mechanical sensitivity and prevented the loss of nerve myelin (Philpott, O'Brien, and McDougall 2017). This suggests that local injection of CBD can reduce inflammation and mechanical sensitivity when administered prophylactically or after the initiation of arthritis.

Two randomized placebo-controlled, veterinarian- and owner-blinded, cross-over studies have evaluated the analgesic effect of CBD oil in osteoarthritic dogs (Gamble et al. 2018; Mejia et al. 2021). CBD oil (2 mg/kg, oral, twice daily, for four weeks) decreased pain and increased locomotor activity in these sixteen dogs, as assessed by veterinarians (Gamble et al. 2018). No side effects were reported by owners; however, bloodwork did show an increase in alkaline phosphatase during CBD treatment. This increase may be attributed to cytochrome P450 mediated oxidative metabolism of the liver, suggesting that monitoring liver enzyme values in pets (and human patients) may be wise. In contrast, CBD oil (2.5 mg/kg, oral, twice daily, for six weeks) did not produce any changes in gait analysis, activity counts (via accelerometry), and pain measures, as indicated by owners (Mejia et al. 2021) The lack of effect in this study may be due to the large variability detected in CBD levels, which ranged from 66 to 860 ng/mL. Adverse events associated with CBD administration included elevation in liver enzymes in fourteen dogs. There are several limitations with these studies, including the short duration of data collection, small sample sizes, lack of a washout period, use of hemp oil as the base for the placebo group, and inconsistent use of other analgesics. Further research is needed.

## CBD and Other Animal Models of Inflammatory Conditions

### Hypoxic Ischemic Brain Injury

Hypoxic ischemic brain injury is the consequence of an insufficient amount of blood flow to the brain, such as after cardiac arrest. This damage can also occur in newborns deprived of oxygen during or following birth. Currently brain damage due to hypoxia ischemia in newborns is treated with hypothermia, but it reduces mortality only in mild cases. Therefore, complementary therapies are necessary to improve patient outcomes. Researchers investigating new treatments for newborn hypoxic ischemic brain injury typically use newborn pigs or rats. Following injury, measures such as changes in brain activity, quantification of brain damage, and pro-inflammatory factors and behavioral performance can be compared in drug-treated versus control animals.

When administered thirty minutes after hypoxic ischemic brain injury in newborn pigs, CBD (1 mg/kg, i.v.) caused recovery of brain activity during the six-hour observation period and reduced the number of damaged neurons in the brain (Pazos et al. 2013). CBD also reduced levels of the cytokine IL-1, suggesting an anti-inflammatory effect (Pazos et al. 2013). All of CBD's beneficial effects were mediated through action at the $CB_2$ receptor (possibly through its action as a partial agonist; Tham et al. 2019) or as a negative allosteric modulator of the $CB_2$ receptor (Martinez-Pinilla et al. 2017) and $5\text{-}HT_{1A}$ receptors (Pazos et al. 2013). In addition, when given fifteen minutes after hypoxic ischemic brain injury, CBD (0.1 mg/kg, i.v.) also restored brain activity, reduced seizures, and reduced brain cell damage in newborn pigs (Alvarez et al. 2008). Furthermore, in a newborn rat model, CBD (1 mg/kg, s.c.) administered ten minutes after hypoxic ischemic brain injury reduced inflammation and the size of brain lesion and improved behavioral performance (Pazos et al. 2012). When combined with hypothermia (the currently approved treatment), CBD (1 mg/kg, i.v.) enhanced the anti-inflammatory effects (reducing the increase in TNF-α), reduced neuronal death, and restored brain activity in newborn pigs (Lafuente et al. 2016; Barata et al. 2019). The therapeutic window for CBD's beneficial effects seems to extend to eighteen hours following hypoxic ischemic brain injury (Mohammed et al. 2017). Taken together, these findings suggest that CBD administration (within eighteen hours after hypoxia ischemia) as an adjunct treatment to hypothermia may be neuroprotective in animal models of hypoxic ischemic brain injury.

**Gastrointestinal Tract Inflammation**
The gut is composed of a selectively permeable barrier, allowing absorption of nutrients and water but preventing the transfer of harmful material such as the inflammatory toxins known as lipopolysaccharides. Inflammation compromises this intestinal barrier, making it more permeable, allowing toxins into the systemic circulation, leading to disease states such as inflammatory bowel disease (IBD). IBD, which includes Crohn's disease and ulcerative colitis, is a chronic, relapsing, and remitting disorder characterized by excessive inflammatory responses in the gastrointestinal tract, leading to damage as well as motility and secretory disturbances. This can result in gastrointestinal bleeding, diarrhea, abdominal

cramping, and malnutrition. IBD is commonly treated with steroids or 5-aminosalicylic acid (5-ASA, mesalazine or mesalamine). It can be managed with medications that maintain remission but are not curative, highlighting the need for new therapies. Interestingly, a retrospective analysis of 298 cannabis users with ulcerative colitis (in comparison to matched noncannabis-using colitis patients) showed lower rates of partial or total surgical removal of the colon and rectum, shorter hospital stays, and a trend toward a lower prevalence of bowel obstruction (Mbachi et al. 2019). These findings suggest that cannabis use may mitigate some of the complications associated with colitis. The role that CBD plays specifically in these benefits remains to be determined.

To determine its anti-inflammatory effects, CBD (30 mg powder, oral) or placebo was given to ten healthy adults, and blood was collected ninety minutes later. This blood was then stimulated with bacterial lipopolysaccharide to induce an inflammatory response. CBD treatment suppressed the pro-inflammatory marker TNF$\alpha$ in these blood samples (Hobbs et al. 2020), suggesting an anti-inflammatory effect of CBD in these healthy adults.

Using a human carcinoma colon cell line (Caco-2 cells) or human colon tissue, researchers can initiate inflammation by treating cells with known pro-inflammatory cytokines such as TNF$\alpha$ and IFN$\gamma$. The degree of inflammation can be assessed by measuring levels of other pro-inflammatory factors and whether these levels are changed by treating the cells with a compound such as CBD. Indeed, CBD (10 µM) prevented the inflammatory response in human colon tissue by reducing the production of pro-inflammatory cytokines (IL-8, IL-6 and MCP-1), an effect mediated by $CB_2$ and TRPV1 receptors (Couch et al. 2017).

To assess gut permeability changes, the administration of ethylenediaminetetraacetic acid to Caco-2 cells induces an increase in permeability, as measured by changes in electrical resistance, an index of the integrity of the cellular barrier in cell culture experiments.

If application of a compound such as CBD can reverse the toxin-induced increase in permeability, this compound could be beneficial in treating the increased intestinal permeability that occurs in IBD. Indeed, CBD (10 μM) accelerated the recovery from the cytokine-induced increased permeability in Caco-2 cells, and this effect was blocked by a $CB_1$ receptor antagonist, but not by a $CB_2$ receptor antagonist (Alhamoruni et al. 2012, 2010; Couch et al. 2019). It is likely that rather than directly acting at the $CB_1$ receptor, CBD is indirectly blocking the increases in gut permeability associated with the endocannabinoids anandamide and 2-AG (Alhamoruni et al. 2010, 2012) by its action as a negative allosteric modulator at the $CB_1$ receptor.

In rodents, IBD can be chemically induced by administration of inflammatory agents such as lipopolysaccharide or sulfonic acid, producing immediate onset of inflammation. When given for three days prior to the inflammatory agent, CBD (5, 10 mg/kg, i.p., daily for six days) reduced colon injury in mice (Borrelli et al. 2009). And when given thirty minutes prior to and for up to four days following the toxin-induced inflammation, CBD (1, 10, 20 mg/kg, i.p., or 20 mg/kg, intrarectally) regulated the inflammatory reaction in the intestine (reduced TNF-α expression and/or lowered IL-6 levels), preventing intestinal damage (De Filippis et al. 2011; Jamontt et al. 2010; Wei et al. 2020; Schicho and Storr 2012). Interestingly, a low dose of CBD (10 mg/kg, i.p.), when combined with a low dose of $\Delta^9$-THC (5 mg/kg, i.p.), enhanced the reduction of colon inflammation and damage in rats (Jamontt et al. 2010). In addition, CBD (1 mg/kg, i.p., administered thirty minutes before colitis induction) reduced intestine disturbance (Wei et al. 2020). Taken together, these findings from cell lines, human colon tissue, and animal studies suggest that CBD may reduce colon inflammation and damage and may restore alterations in intestine permeability, all symptoms of irritable bowel disease.

**CBD and Irritable Bowel Disease Patients**
Five studies have evaluated the effect of CBD on intestinal permeability or inflammation in humans. First, a randomized, double-blind, placebo-controlled trial examined the effect of CBD (600 mg, oral) on the excretion of sugar probes (as an in vivo measure of intestinal permeability) after aspirin-induced inflammation in thirty healthy volunteers (Couch et al. 2019). CBD prevented the aspirin-induced increase in intestinal permeability, suggesting that CBD may restore intestinal permeability and may be beneficial in IBD patients.

Four clinical trials have assessed the efficacy of CBD in IBD patients. A randomized, placebo-controlled trial in twenty-five Crohn's disease patients showed that CBD (10 mg, oral, twice daily for eight weeks) did not improve disease (Naftali et al. 2017), but the rather low dose of CBD used in this study may not have been sufficient to overcome this chronic inflammation. Indeed, a randomized, double-blind, placebo-controlled pilot study with sixty IBD patients showed that treatment with a higher dose of CBD (50 mg, oral capsule, twice daily for ten weeks) improved illness severity and patient-reported quality-of-life outcomes, although no difference was found in rates of remission between the CBD and placebo groups (Irving et al. 2018). Another randomized, double-blind, placebo-controlled, cross-over trial found no effect of CBD chewing gum (50 mg CBD per piece of gum, self-titrating up to six pieces per day for eight weeks) on pain scores in thirty-two irritable bowel disease patients (van Orten-Luiten et al. 2021). Large within- and between-group differences such as the number of gums used and subjectivity in self-report pain severity may have masked any treatment effects. Most recently, a double-blind, randomized, placebo-controlled, single-center trial evaluated CBD-rich cannabis oil (160:40 mg/ml CBD:THC) for eight weeks in fifty-six Crohn's disease patients (Naftali et al. 2021). Those treated with the CBD-rich cannabis oil showed clinical improvements in general well-being, abdominal pain, and quality-of-life measures.

The inclusion of CBD plasma concentrations in some of these studies would be beneficial to determine effective dosing. Although these initial studies do suggest that a higher dose of CBD may provide therapeutic benefit for IBD patients, further investigation of optimal dosing and extended durations of treatment is necessary.

## CBD and Pain: Human Studies

The Canadian Pain Society lists cannabinoids as fourth-line analgesics in their guidelines for the management of chronic neuropathic pain (Moulin et al. 2007). This means that other recommended analgesics would need to demonstrate ineffectiveness before cannabinoids could be prescribed. Survey reports suggest that patients are using cannabis to manage their pain, with approximately a 50 percent reduction in self-reported pain (Cuttler, LaFrance, and Craft 2020). An electronic survey from 2,032 medicinal cannabis patients revealed that pain syndromes accounted for 42.4 percent of the illnesses being treated, with chronic pain being the most common reason for cannabis use (Baron et al. 2018). Furthermore, many of these pain patients substituted cannabis for their opiate medications (40.5–72.8 percent). A historical study of 37 chronic pain patients using opiates showed that those using cannabis were more likely to reduce their daily opioid prescription dosages during the study or to stop filling their opioid prescriptions completely (Vigil et al. 2017). A report from the National Academy of Sciences (2017) indicated substantial evidence that cannabis is an effective treatment for chronic pain in adults, but this conclusion was mostly based on studies investigating nabiximols in pain patients.

Nabiximols (Sativex)—a highly standardized oromucosal spray containing 2.7 mg THC and 2.5 mg CBD—was approved by Health Canada in June 2005 for prescription for multiple sclerosis patients with central neuropathic pain. In August 2007, Sativex was also approved for the treatment of cancer pain unresponsive to opioid therapy. Nabiximols has been evaluated in multiple clinical trials for pain management (reviewed below). It is important to determine the role that CBD alone plays in the pain and inflammation improvements reported by cannabis and nabiximols users. Indeed, very little is known about the efficacy, dose, routes of administration, or side effects of CBD products.

An audit on the first 400 patients in New Zealand prescribed CBD oil showed significant improvements in self-reported pain (Gulbransen, Xu, and Arroll 2020), and a survey of 58 patients in a palliative medicine practice in Florida showed that 24 percent of patients reported using CBD, with 50 percent of these patients reporting pain relief by CBD (Highet et al. 2020). Survey data also suggest that chronic pain patients using CBD report pain relief, and at least 53 percent of them have reduced or ceased their opiate use (Schilling et al. 2021; Boehnke et al. 2021; Capano, Weaver, and Burkman 2020).

Despite these positive anecdotal reports of CBD's pain-relieving effects, few properly controlled studies have investigated the effects of CBD in human patients. Some case studies have investigated CBD's ability to reduce pain. In a recent case series with seven kidney transplant patients treated with CBD (50 to 150 mg twice a day for three weeks) for chronic pain, six out of seven patients reported an improvement in pain (Cunetti et al. 2018). CBD was well tolerated, without severe adverse effects. Two opioid-naive patient case reports (Eskander et al. 2020) suggested improved lower back pain due to a fall or surgery with the use of topical CBD cream (independently validated to contain CBD). In a retrospective, observational, open-label study, sublingual CBD-rich hemp oil (25 mg/kg daily, to a maximum dose of 150 mg/ml, over three months) improved body pain in twelve females with adverse drug effects due to the human papillomavirus (HPV) vaccine (Palmieri, Laurino, and Vadala 2017).

CBD alone has been investigated in a few published clinical trials. The first, a double-blind, randomized, placebo-controlled, single-patient cross-over trial with two-week treatment periods, involved twenty pain patients unresponsive to standard treatment (Wade et al. 2003). Most were multiple sclerosis patients with symptoms of neuropathic pain and spasticity. CBD (average dose of 22.5 mg/day, sublingual spray) reduced pain and spasticity scores over the two-week treatment period. CBD was well tolerated; in the CBD group, 33 percent of patients reported an adverse event (versus 48 percent of patients in the placebo group)—nausea, headache, and diarrhea, for example. Although these findings may seem promising, the sample in this study was very small, and patients were able to self-titrate, making establishing an effective dose difficult. Future studies should address this dosing issue and investigate CBD's effectiveness in longer treatment durations. Importantly, a crossover placebo design study suggests that study participant expectations play a large role in CBD's analgesic effect (De Vita et al. 2021). Healthy participants reported more pain reduction if they were told that they received CBD, regardless of actual drug content.

A handful of studies suggest that CBD may not be effective in reducing pain in healthy volunteers or pain patients. A double-blind, placebo-controlled, within-subject study evaluated the analgesic effects of CBD (200–800 mg, oral) in seventeen healthy volunteers (Arout et al. 2021). CBD did not modify pain threshold or tolerance and all doses of CBD increased ratings of painfulness. These findings suggest that acute administration of CBD does not have an analgesic effect in healthy participants. Similarly, a randomized, placebo-controlled, double-blind crossover study failed to show a significant effect of CBD (800 mg, oral) on acute pain in healthy participants (Schneider et al. 2022). A single-center, randomized, double-blind, placebo-controlled study showed no benefit of CBD (20–30 mg oral capsule daily for twelve weeks) over placebo for pain relief in 129 patients with hand osteoarthritis and psoriatic arthritis (Vela et al. 2021). In addition, a single-center, randomized, double-blind, placebo-controlled study evaluated oral CBD (400 mg, oil) as an adjunct treatment for pain reduction in 100 patients who presented with acute, nontraumatic low back pain (Bebee et al. 2021). There were no differences between CBD and placebo for pain scores assessed two hours after treatment, length of stay, need for rescue analgesia, or adverse events. These findings suggest no superior acute analgesic effects for CBD for relieving acute pain. Because no serum CBD levels were measured, it is difficult to determine whether therapeutic blood concentrations were achieved in these patients at the time of assessment.

Two studies have used topical application of CBD for pain relief. The first, a double-blind, placebo-controlled study evaluated the efficacy of transdermal CBD in sixty patients with myofascial pain (face and neck pain and headache due to muscle damage; Nitecka-Buchta et al. 2019). CBD treatment (ointment with 20 percent CBD oil, made from oil with 70 mg CBD/ml, twice daily for fourteen days) decreased pain intensity ratings, and no adverse events were reported (Nitecka-Buchta et al. 2019). Similarly, a four-week, placebo-controlled, cross-over trial evaluated topically delivered CBD oil in twenty-nine neuropathic pain patients (Xu, Cullen et al. 2020). CBD (250 mg CBD/3 fl. oz for four weeks) reduced pain ratings, and no adverse events were reported (Xu, Cullen et al. 2020). Taken together, these findings suggest that topical administration of CBD oil may reduce myofascial pain and neuropathic pain without adverse effects and thus may be an effective treatment, devoid of the adverse effects often reported with other pain management medications.

In 2021, consensus-based recommendations on how to dose and administer medical cannabis to chronic pain patients were developed by global experts (Bhaskar et al. 2021). They developed three treatment protocols: routine, conservative, or rapid. A *routine* treatment involves starting a patient with a CBD-predominant product at a dose of 5 mg twice daily and titrating the CBD-predominant product by 10 mg every two to three days until the patient reaches their goals, or up to 40 mg/day. At a CBD-predominant product dosage of 40 mg/day, clinicians may consider adding THC at 2.5 mg and titrate by 2.5 mg every two to seven days until a maximum daily dose of 40 mg/day of THC. A *conservative* protocol involves starting the patient on a CBD-predominant product at a dose of 5 mg once daily and titrating the CBD-predominant dose by 10 mg every two to three days until the patient reaches their goals, or up to 40 mg/day. At a CBD-predominant dose of 40 mg/day, clinicians may consider adding THC at 1 mg/day and titrate by 1 mg every seven days until a maximum daily dose of 40 mg/day of THC. A *rapid* protocol involves starting the patient on a balanced THC/CBD product at 2.5–5 mg of each cannabinoid once or twice daily and titrating by 2.5–5 mg of each cannabinoid every two to three days until the patient reaches their goals, or to a maximum THC dose of 40 mg/day.

Consensus-based recommendations have been also developed for the use of cannabis in chronic pain patients who are not achieving pain management with opioids (Sihota et al. 2020). There was consensus observed in the initiation of use with an oral CBD extract in the daytime with consideration of adding a low dose of THC. Further, opioid tapering was recommended to be initiated once the patient has reported improved function, seeks less pain control medication, or has optimized his or her cannabis dose.

## Nabiximols and Pain: Human Studies

Although few studies have evaluated the efficacy of CBD treatment in pain patients, nabiximols (a sublingual spray with approximately a 1:1 ratio of THC:CBD) has been evaluated in multiple clinical trials for pain management for a number of different pain conditions, with generally favorable results. Unfortunately, because nabiximols is a combination of THC and CBD, we cannot know whether the pain-relieving effects are mediated by THC or CBD, or both.

## Multiple Sclerosis Patients

In multiple sclerosis patients, nabiximols has been shown to reduce spasticity (muscle stiffness and spasms; Vermersch and Trojano 2016; Haupts et al. 2016; Trojano and Vila 2015; Flachenecker, Henze, and Zettl 2014; Patti et al. 2016; Serpell, Notcutt, and Collin 2013; Notcutt et al. 2012; Meuth et al. 2020) and pain (Markova et al. 2019; Paolicelli et al. 2016; Langford et al. 2013; Ferre et al; 2016, Wade et al. 2003, 2004; Zajicek et al. 2003; Wade et al. 2003; Rog et al. 2005; Turri et al. 2018) when given as an adjunct therapy. As these improvements were shown for at least fourteen weeks of treatment (Ferre et al. 2016), these findings suggest that CBD may offer long-term pain reduction in MS patients who are unresponsive to typical analgesics.

## Cancer Pain Patients

Nabiximols has also been investigated for pain relief in cancer patients unresponsive to opioid analgesics. (Also refer to chapter 5 for a discussion of CBD's effects in cancer and chapter 6 for a discussion of CBD's effects in chemotherapy-induced nausea and vomiting.) Although some studies have shown pain relief in cancer patients using nabiximols as an adjunct treatment (Portenoy et al. 2012, Johnson et al. 2010), some have shown no such pain relief (Johnson et al. 2013; Lichtman et al. 2018; Fallon et al. 2017). These differential results may be due to differing nabiximols dosing regimens, differing efficacy of adjunct treatments, or differing disease severity or progression. Specifically, in a study by Johnson et al. (2013) in terminal cancer patients whose pain was not fully relieved by the current strong opioid analgesics, nabiximols reduced their pain, and over the five-week study, patients did not need to increase their dose of nabiximols or other pain medications (Johnson et al. 2013). These findings suggest that the adjunct use of nabiximols in cancer-related pain may be beneficial for some treatment-resistant patients.

### Neuropathic Pain Patients

Nabiximols has also been investigated in the treatment of neuropathic pain patients. Nabiximols as an adjunct treatment reduced pain and improved allodynia (Nurmikko et al. 2007; Serpell et al. 2014; Hoggart et al. 2015; Ueberall, Essner, and Mueller-Schwefe 2019). In addition, patients using nabiximols required fewer tablets of rescue medication (Serpell et al. 2014). Notably, in a study of neuropathic pain patients, adjunct treatment with oral THC (19 percent) and CBD (less than 1 percent) solution, with a mean dose of 68.5 mg/day for twelve months, reduced pain, an effect maintained during the entire year of treatment, with minimal dosage titration (Mondello et al. 2018). These findings suggest that the adjunct use of nabiximols by neuropathic pain patients may be beneficial and may offer long-term efficacy.

### Arthritis Patients

In a randomized double-blind trial (Blake et al. 2006) with fifty-eight rheumatoid arthritis patients, nabiximols (average dose at week 5, 14.6 mg THC:13.5 mg CBD, oral spray) reduced pain during movement and pain at rest with no serious adverse events reported over five weeks. These improvements by CBD are noteworthy, as they were seen in a group of rheumatoid arthritis patients for whom standard medication was ineffective.

### Postoperative Pain Patients

A prospective study analyzed the use of CBD/THC products (containing CBD and/or THC, including topical creams, tinctures, smoking cannabis, and edibles) in the perioperative period for 195 hip and knee reconstruction or replacement patients (Runner et al. 2020). No differences were found between CBD/THC users and nonusers in the length of opiate use, total morphine milligram equivalents taken, opiate pills taken, average post-op pain scores, the percentage of patients requiring a refill of opiates, or length of stay. Similarly,

Jennings et al. (2019) retrospectively reviewed 71 patients who underwent knee reconstruction/replacement who self-reported cannabis use, and found no difference in length of stay, in-hospital total morphine equivalents, or postoperative range of motion, suggesting that cannabis use does not seem to alter short-term outcomes in patients undergoing total knee arthroplasty. But a prospective, randomized study of 73 patients undergoing elective operations (Jefferson et al. 2013) revealed that cannabis users needed more rescue analgesics during recovery (within the first six hours after general anesthesia) and had greater pain intensity scores in the first hour after surgery compared to nonusers. This suggests that cannabis users undergoing surgery may require more opioid rescue analgesia in the early postoperative period, but a history of cannabis use does not seem to alter the recovery outcomes of patients following surgery.

Unfortunately, the mechanism responsible for the apparent opioid-tolerance-producing effects in cannabis users is still not fully understood. Chronic cannabis use (or THC administration) does lead to the downregulation and desensitization of $CB_1$ receptors and alterations in receptor binding (Villares 2007; Oviedo, Glowa, and Herkenham 1993; Romero et al. 1998; Sim-Selley 2003), which would result in a reduced analgesic effect of the endogenous cannabinoid system and perhaps the opioid system. Indeed, chronic administration of THC in rodents has been shown to induce tolerance to the analgesic effect of opioids (Smith, Welch, and Martin 1994; Welch 1997), although some other studies have not shown this effect (Mao et al. 2000), and some have even found enhanced analgesic effects with THC administration in morphine-tolerant rats (Rubino et al. 1997). Longitudinal studies are needed to determine if cannabis use causes alterations in the pain system or if these individuals begin to use cannabis to "treat" their altered system.

**Fibromyalgia Patients**

A randomized placebo-controlled four-way crossover trial (van de Donk et al. 2019), explored the analgesic effects of inhaled pharmaceutical-grade cannabis in twenty chronic pain patients with fibromyalgia. Patients inhaling a strain with a known higher THC and CBD content (13.4 mg THC, 17.8 mg CBD) displayed a 30 percent decrease in pain scores compared to placebo, suggesting a potential beneficial effect for these patients consuming this high THC and CBD strain of cannabis. Patients inhaling a cannabis variety with a high CBD and very low THC content (18.4 mg CBD and less than 1 mg THC) did not demonstrate any improvements in pain compared to placebo. Because this study involved only a single CBD administration, this may be insufficient to improve pain scores or this dose may be too low.

## Conclusion

A large proportion of medicinal cannabis patients are using cannabis to treat pain syndromes, particularly for chronic pain alleviation. Cannabis use for unmanageable pain has been shown to reduce the dose of opiates required to achieve pain control, suggesting opioid-sparing effects (O'Connell et al. 2019). Although no published clinical trials have clearly demonstrated this opioid-sparing effect with CBD alone, according to www.clinicaltrials.gov, registered clinical trials are investigating the interaction of CBD and morphine effects on pain sensitivity (NCT04030442) and abuse liability (NCT03679949).

Animal models suggest that CBD is generally ineffective in reducing acute pain when administered systemically, but this may be because insufficient amounts of CBD are reaching the brain to produce an analgesic effect. When CBD is combined with THC or morphine, it can potentiate the analgesic effects of these compounds to relieve acute pain, and analogs of CBD also seem to have analgesic effects in acute pain models in animals.

In animal models of neuropathic pain, CBD is effective in preventing the development of certain types of chemotherapy- and diabetes-induced pain, as well as the treatment of already established chemotherapy-, spinal cord injury-, or postoperative-induced neuropathic pain. Indeed, a clinical study with topical CBD improved pain ratings in neuropathic pain patients. Nabiximols has also been shown to reduce pain and allodynia in neuropathic pain patients, with improvements maintained for an entire year.

In animal models, when administered prior to or following the initiation of inflammation or arthritis, CBD can abolish neuropathic pain symptoms and reduce the associated swelling. Nabiximols has also been shown to reduce pain in rheumatoid arthritis patients who were unresponsive to standard treatments. These findings suggest that CBD may be an adjunct treatment to investigate in inflammatory pain conditions such as arthritis.

CBD's anti-inflammatory effects have also been demonstrated in animal models of brain injury due to interrupted blood flow to the brain. CBD treatment within eighteen hours can reduce brain cell damage and inflammation and restore brain activity. In animal and cellular models of gastrointestinal tract inflammation, CBD has anti-inflammatory effects and restores the integrity of the gut lining. A handful of human studies also suggest that CBD may improve inflammatory bowel disease symptoms and restore intestinal integrity, but further research is needed.

Sublingual spray-delivered CBD alone (as well as nabiximols) reduced pain and spasticity scores in multiple sclerosis patients, and topical application of CBD relieved myofascial pain and neuropathic pain. To date, nearly fifty clinical trials registered at www.clinicaltrials.gov are investigating the ability of CBD (alone or in combination with THC) to reduce pain associated with conditions such as tooth extractions, chronic back pain, postsurgical pain, diabetic neuropathy, and cancer pain. These important properly controlled randomized clinical trials will help to better understand the potential analgesic properties of CBD.

# 8
# Anxiety

Although the emotions of fear and anxiety are not only adaptive but essential for survival, when experienced in excess or continually, these responses become maladaptive and can lead to the development of neuropsychiatric disorder(s), such as generalized anxiety disorder, social anxiety disorder, and posttraumatic stress disorder (PTSD, discussed in chapter 9). Anxiety disorders are typically treated with pharmacotherapy or psychological therapy, or both. The most common psychological therapy is cognitive behavioral therapy (CBT). CBT has demonstrated its usefulness in the treatment of anxiety disorders (Kaczkurkin and Foa 2015), but it can often be difficult to implement due to limited resources (Bandelow et al. 2007). First-line pharmacotherapies include selective serotonin reuptake inhibitors (SSRIs) and serotonin-norepinephrine reuptake inhibitors (SNRIs), and other treatment options include pregabalin, tricyclic antidepressants, buspirone, moclobemide, and benzodiazepines (see Bandelow, Michaelis, and Wedekind 2017 for a review). Despite these available treatments, some 40 to 60 percent of patients continue to have residual symptoms, do not adhere to the treatment, or have difficulty accessing treatments (Katzman et al. 2014). Indeed, these pharmacotherapies can have undesirable side effects such as risk of dependence and withdrawal syndrome, sexual dysfunction, cognitive and psychomotor impairment, insomnia, and weight gain (Katzman et al. 2014). These treatment challenges highlight the need to evaluate new treatments (or treatment adjuncts) for these disorders.

Cannabis use is common in those with anxiety disorders, although the nature of the association between cannabis use and anxiety disorders remains unclear (see Halladay et al. 2020 for review). A meta-analysis of longitudinal studies (Kedzior and Laeber 2014) found a small, significant association between early cannabis use and later anxiety in the general population (odds ratio = 1.15), but with the inclusion of only high-quality studies, this relationship became nonsignificant (Twomey 2017). This suggests there may be an association between cannabis use and anxiety disorders and that cannabis use may precede anxiety disorder development, but further high-quality studies are needed to fully understand this association. This link between cannabis use and anxiety may arise because of brain changes that occur due to early use of cannabis, or that those with anxiety disorders may be predisposed to use cannabis more frequently, or it may be that those with anxiety choose to self-medicate.

A meta-analytic review (Kosiba, Maisto, and Ditre 2019) of survey outcomes of reported cannabis use suggests that cannabis is being used to improve anxiety by up to 50 percent of those surveyed. Cannabis is often used for antianxiety purposes, yet *increased* anxiety and panic reactions are reasons often reported for cessation of cannabis use (Thomas 1993; Reilly et al. 1998; Schofield et al. 2006). The anxiety-relieving versus anxiety-promoting effects of cannabis may be a function of the varying concentrations of cannabinoids within a given cannabis strain and how these ratios interact to ultimately produce their effect. In addition, these variable effects of cannabis could be caused by a downregulation of the endogenous cannabinoid system due to repeated cannabis exposure (Hirvonen et al. 2012), which could ultimately cause tolerance to the antianxiety effects of cannabis. Indeed, both THC and CBD have been found to have a dose-dependent effect on anxiety, but usually in opposing directions. In general, THC produces or increases anxiety, at high doses, while some evidence indicates that CBD may reduce anxiety at moderate doses and under certain experimental conditions (see Crippa et al. 2009 for review). Survey data and self-reports indicate that CBD users regard this substance as a means to manage anxiety (Corroon and Phillips 2018; Wheeler et al. 2020; Tran and Kavuluru 2020; Leas et al. 2020), although CBD's therapeutic benefit for anxiety has not been supported to date by findings from randomized clinical trials.

Also of note is the potential for drug-drug interactions in those using pharmacotherapies to treat their anxiety disorder. CBD undergoes hepatic metabolism via the cytochrome P450 enzymes, including CYP2C34A and CYP2C19. Because this P450 enzyme is responsible for the metabolism of many commonly prescribed medications, CBD may interact with other CYP substrates to produce adverse reactions. CBD may increase the concentrations of other CYP3A4 substrates, such as the benzodiazepines alprazolam and triazolam, as well as CYP2C19 substrates, such as the antidepressants citalopram and fluoxetine. Patients taking these types of medications may be at risk of increased adverse effects and may require lower doses or more frequent monitoring, or both.

## CBD's Preclinical Effects on Anxiety

CBD's ability to reduce anxiety has been evaluated in several animal models of general anxiety, such as the elevated plus maze and the light-dark emergence test. The findings from CBD's effects in these paradigms are discussed below.

The elevated plus maze (EPM) was developed from the work of Montgomery (1955) and relies on rodents' inherent tendency to explore a novel place but also to avoid the aversive aspects of that novel place. This maze consists of two open and two enclosed arms. The animal is placed in the center of the maze and tracked as it explores the maze for a short period of time (usually five minutes). Rats instinctively spend more time exploring an enclosed, safe space versus an open one. Because they tend to avoid open spaces, the amount of time they spend in the open arms of the maze is used as an index of anxiety: animals that are less anxious (displaying antianxiety behavior) will spend more time in this open, more anxiety-provoking arm (relative to controls). Indeed, higher levels of corticosterone (the main stress hormone in rodents) are measured in the blood of animals confined to the open arms versus the closed arms of the maze (Pellow et al. 1985). Furthermore, antianxiety behavior has been observed in rats and mice treated with classic antianxiety treatments such as benzodiazepines and barbiturates. In general, using the EPM, CBD has been shown to exert antianxiety effects in rodents at moderate doses and is ineffective (but does not increase anxiety) at higher doses (see Blessing et al. 2015 for an excellent review). CBD's antianxiety effects at moderate doses are similar to those seen with the antianxiety compound diazepam in mice (Onaivi, Green, and Martin 1990). In addition, the fluorination of CBD enhances its antianxiety effects in the EPM (Breuer et al. 2016).

Another rodent anxiety test, the light-dark emergence test, originally described by Crawley and Goodwin (1980), consists of two chambers—one large and brightly lit and the other smaller and dark. The light serves as an anxiety-provoking stimulus, establishing a conflict between the rodents' desire to explore the novel open compartment and their desire to avoid the brightly lit, open chamber. Drugs that increase the amount of time spent in the brightly lit chamber (relative to controls) are thought to have antianxiety effects. This test provides a low-level baseline of anxiety responding and therefore may be more sensitive in detecting both increases and decreases in anxiety behavior (Holmes 2001; Bourin and Hascoet 2003). Indeed, several benzodiazepines have antianxiety effects in this test, and CBD also shows antianxiety effects in rodents at low to moderate doses and is ineffective (but does not produce anxiety) at higher doses (Long et al. 2010; Guimaraes et al. 1990). In addition, CBD seems to maintain its antianxiety effects when chronically (twenty-one days) administered (Long et al. 2010).

Differential effects of CBD on anxiety have been reported under low-stress and high-stress conditions. Restraint stress can increase subsequent anxiety behaviors in rats, which can be ameliorated by CBD (10 mg/kg, i.p.; Resstel et al. 2009). Similarly, Song et al. (2016) found that CBD (10 mg/kg, i.p.) enhances contextual fear memory extinction under high-fear conditions versus low-fear conditions. The light-dark emergence test can also be utilized to investigate the stress-induced anxiety response by exposing animals to a stressor such as foot shock and then testing them twenty-four hours later (Bluett et al. 2014). Reports suggest that animals treated with CBD (2.5, 5 mg/kg, i.p.) did not differ from controls when tested under low-stress conditions (Rock et al. 2017; O'Brien, Wills et al. 2013), suggesting that in low stress, CBD did not reduce (or increase) anxiety behavior. Most interesting, when tested under high-stress conditions (foot shock twenty-four hours before the test), CBD (5 mg/kg, i.p.) had antianxiety effects (Rock et al. 2017). Similarly, the acidic precursor of CBD, cannabidiolic acid (CBDA, 0.1–100 µg/kg, i.p.), when administered acutely and chronically (twenty-one days), was ineffective in low-stress conditions (Brierley et al. 2016; Rock et al. 2017), but when tested under high-stress conditions, CBDA had antianxiety effects (Rock et al. 2017). In addition, the methyl ester version of CBDA, HU-580 (Pertwee et al. 2018), also exhibits potent stress-induced antianxiety effects in the light-dark test. The results of these studies suggest that CBD and CBDA may be highly effective treatments for the reduction of anxiety but only among individuals in a high state of stress.

The mechanism by which CBD exerts its antianxiety effects in rodents is still under investigation, in part because CBD can act on several pathways that may modulate anxiety. CBD has been described as a negative allosteric modulator of the $CB_1$ receptor (Laprairie et al. 2015), which may be the mechanism underlying its ability to attenuate THC's pro-anxiety effects. Several of CBD's antianxiety effects seem to be driven by its action at the $5\text{-}HT_{1A}$ receptor. Indeed, when a selective $5\text{-}HT_{1A}$ receptor antagonist is delivered to block this receptor, CBD's antianxiety effects are also blocked, indicating that this receptor is necessary for CBD to reduce anxiety (Campos and Guimaraes 2008). This mechanism of action for CBD's antianxiety effect is reasonable, given that other known $5\text{-}HT_{1A}$ receptor agonists (for example, buspirone) are approved treatments for general anxiety disorder. The ineffectiveness of high doses of CBD in anxiety models may be due to CBD's action at TRPV1 receptors because blockade of these receptors restores CBD's antianxiety effects (Campos and Guimaraes 2009).

Together, these preclinical studies suggest that CBD may have antianxiety-like effects (especially in pre-stressed animals) when given systemically (both acutely and chronically) or infused locally into various brain areas important in fear and anxiety (see Lee et al. 2017 for an excellent review). In general, CBD at doses of 0.5 to 10 mg/kg (i.p.) in mice (Onaivi, Green, and Martin 1990) and 2.5 to 10 mg/kg (i.p.) in rats (Guimaraes et al. 1990) produces antianxiety-like effects. A single study examining the effect of prenatal CBD (20 mg/kg CBD, orally, daily from two weeks prior to mating through gestation and lactation) indicated that offspring exposed to CBD during development exhibited increased anxiety (Wanner et al. 2021). Indeed, the effect that CBD may have on development has not been properly explored.

## CBD's Effects on Anxiety in Humans

In examining the available studies assessing the effects of CBD on anxiety, the National Academies of Sciences report (2017) indicates limited evidence that CBD improves anxiety symptoms due to limitations in these studies. Such limitations include randomization procedures, single doses of CBD given (in some cases), absence of plasma CBD concentrations, and whether results are applicable to all anxiety disorders. These caveats must be kept in mind when interpreting the human studies presented in this chapter.

A handful of studies exploring the antianxiety effects of CBD have been conducted with human participants. The first of these studies focused on CBD's ability to reduce the anxiety that THC can produce in healthy participants. Although the use of healthy participants might provide future direction for dosing, CBD's efficacy must ultimately be demonstrated in those with an anxiety disorder diagnosis if it is to be clinically meaningful.

The first (Karniol et al. 1974) double-blind, placebo-controlled study of forty healthy male volunteers showed that CBD (15, 30, 60 mg, orally in ethanol and orange juice) reduced the anxiety produced by THC (30 mg, orally in ethanol and orange juice). Subsequently, in a double-blind, placebo-controlled, crossover study conducted on eight healthy volunteers, CBD (1 mg/kg, orally in ethanol and lemon juice) reduced the anxiety that was produced by THC (0.5 mg/kg, orally in ethanol and lemon juice; Zuardi et al. 1982). These early results suggest that the anxiety produced by higher doses of THC could be lessened by CBD. In addition, cannabis strains with less THC and more CBD content could be more likely to reduce anxiety.

Subsequent studies with CBD have induced anxiety with the simulated public speaking (SPS) test, a classic task used to assess the efficacy of antianxiety drugs. In the SPS test, participants are asked to prepare and present a speech that will be delivered to an audience or videorecorded, inducing physiological and subjective changes characteristic of anxiety. Self-report measures such as the visual analogue mood scale (VAMS), the state-trait anxiety inventory (STAI), and a bodily symptoms scale (BSS) are assessed, often by comparing these scores pre- or post-SPS test in those given CBD.

### Effects in Healthy Participants

In a double-blind, placebo-controlled study conducted on forty healthy volunteers exposed to the SPS test, CBD (300 mg, oral capsule) decreased anxiety (Zuardi et al. 1993). In a follow-up study, Zuardi et al. (2017) examined a wider CBD dose range using the SPS test in sixty healthy volunteers and found that relative to placebo, CBD produced antianxiety effects in a dose-dependent, inverted U-shaped curve, such that a moderate dose of CBD (300 mg, oral capsule) reduced anxiety measures, but lower (100 mg, oral capsule) and higher doses (900 mg, oral capsule) were ineffective. This finding was later reproduced by this group, with CBD (300 mg, oral capsule) again reducing anxiety relative to placebo but lower (150 mg, oral capsule) and higher doses (600 mg, oral capsule) having no effect in fifty-seven healthy participants (Linares et al. 2019). These findings suggest a rather narrow therapeutic window of 300 mg for CBD's antianxiety effects in dealing with an anxiety-provoking event in healthy volunteers.

In contrast, a randomized, double-blind study (Hundal et al. 2018) conducted on thirty-two healthy individuals preselected for high trait paranoia induced anxiety using virtual reality by recreating the experience of being on a London underground train. With this procedure CBD (600 mg, oral capsule) had no beneficial effects on anxiety. This may be due to the dose being too high; however, in agreement with this finding, a within-subjects, placebo-controlled, double-blind study found that CBD (300, 600, 900 mg, oral capsule) did not dampen responses to negative emotional stimuli in healthy volunteers (Arndt and de Wit 2017). These studies did not utilize the SPS test, the latter evaluating participants' responses to images or words with negative affective content, their reactivity to threatening emotional faces, and their sensitivity to social rejection. Thus, it is possible that the stimuli used in these studies were less salient than the SPS test and therefore may not have been sufficient to induce anxiety. Alternatively, it is possible that the antianxiety effects of CBD in healthy participants can be revealed only under high levels of anxiety. Clearly the preclinical studies reviewed above suggest that CBD is most effective against anxiety in animals that are prestressed, suggesting that a high baseline level of stress (such as in posttraumatic stress disorder) may be a necessary experimental condition to reveal the antianxiety effects of CBD.

A randomized clinical trial of 120 frontline health care workers caring for COVID-19 patients showed improved anxiety, depression, and emotional exhaustion scores in those taking oral CBD (150 mg twice per day for four weeks) along with standard care, compared with those only receiving standard care (Crippa et al. 2021). Five participants who received CBD with standard care experienced serious adverse events. The interpretation of this study's findings was limited by its single dose and single-center design, short follow-up period, and lack of a double-blind, placebo-controlled environment.

Importantly, a randomized crossover study by Spinella and colleagues (2021) demonstrated CBD's expectancy effect (also known as the placebo effect) in forty-three healthy adults subjected to the Maastricht Acute Stress Test (inducing stress and state anxiety). They underwent two experimental sessions in which they self-administered CBD-free hempseed oil sublingually, but on one of the sessions, they were incorrectly told that the oil contained CBD (the expect CBD condition), while in the other session, they were told that the oil did not contain CBD (the expect no CBD condition). Subjects with the strongest beliefs about CBD's antianxiety effects reported the lowest levels of anxiety when they expected CBD oil and the highest levels of anxiety when they expected CBD-free oil. There were no differences in anxiety measures in those who held low to moderate beliefs about CBD. These findings clearly demonstrate the power of expectancy in modifying the perceived effect of CBD and should be considered when examining CBD's effects in human trials.

**Effects in Participants with Anxiety Disorders** To date, three published studies have examined the antianxiety effects of CBD in those diagnosed with anxiety disorders. In a double-blind, placebo-controlled study conducted on twenty-four healthy participants with untreated social anxiety disorder exposed to the SPS test, CBD (600 mg, oral capsule) reduced anxiety, restoring the participants' anxiety scores to levels seen in healthy controls (Bergamaschi, Queiroz, Chagas et al. 2011). More recently, a double-blind, placebo-controlled study examining the effects of CBD in thirty-seven Japanese teenagers with social anxiety disorder showed that CBD (300 mg, oral oil, daily for four weeks) decreased anxiety (Masataka 2019). In addition, a large, retrospective case series was conducted at a psychiatric clinic involving clinical application of CBD (25–175 mg daily for three months, oral capsule) as an adjunct treatment for forty-seven patients diagnosed with an anxiety disorder. In the first month of treatment, CBD decreased anxiety scores in 80 percent of patients, and at the three-month follow-up, anxiety scores remained decreased (Shannon et al. 2019). Due to small sample size and lack of placebo control (in the retrospective case series), the results of these studies should be interpreted with caution.

A number of clinical trials (www.clinicaltrails.gov) involving CBD as a treatment for anxiety are underway. Two trials (NCT02548559 and NCT04286594) are evaluating the effects of oral spray CBD (10 mg administered 3 times per day for four weeks or 15 mg twice daily for six weeks) for the treatment of anxiety in adults. Another clinical trial (NCT03549819) is examining CBD (200–800 mg per day, oral capsule) for the treatment of adults diagnosed with an anxiety disorder. Finally, a new pilot trial (NCT04267679) plans to investigate the effect of CBD (25 mg up to a total dosage of 100 mg per day for twelve weeks, oral capsule) on individuals with diagnosed anxiety. Indeed, these clinical trials will help to understand the consequences of longer-term CBD administration in patients with anxiety disorders.

**Effects in Psychiatric Clinic Patients** Only one study to date has examined CBD's effects on anxiety in psychiatric clinic patients. A randomized, placebo-controlled, double-blind study (Appiah-Kusi et al. 2020) investigated whether CBD would normalize the neuroendocrine and anxiety responses to public speaking stress in thirty-two clinical patients at a high risk for psychosis. Under placebo conditions, those patients displayed abnormal neuroendocrine and psychological responses to SPS compared with healthy participants, but seven-day treatment with CBD (600 mg, oral capsule) partially restored these altered responses to SPS in these participants at high risk for psychosis. This study suggests that CBD may reduce anxiety in psychiatric clinic patients at high risk for psychosis and restore aberrant neuroendocrine responses, initial findings that support the need for future studies.

**Effects in Parkinson's Disease Patients** Recently a randomized, double-blinded, placebo-controlled, crossover clinical trial showed that CBD (300 mg, oral capsule) decreased anxiety signs and symptoms in twenty-four patients with Parkinson's disease undergoing the SPS test (de Faria et al. 2020). This finding suggests that CBD may be beneficial in helping Parkinson's disease patients cope with anxiety-provoking events, but the efficacy of chronic CBD in this population is unknown. Again, the small sample size is a limitation in interpreting these findings.

**Mechanism of Action in Humans** Few studies have endeavored to investigate how CBD may be acting in the brain to reduce anxiety in humans. Previous research has determined that critical brain regions involved in anxiety are limbic and paralimbic areas such as the orbitofrontal, cingulate and medial temporal cortex, and the insula. Researchers have therefore used imaging techniques to determine how CBD may alter activation of these known anxiety-related brain regions.

A double-blind, placebo-controlled cross-over study in ten healthy volunteers compared the effects of CBD (400 mg, oral capsule) and placebo on resting cerebral regional blood flow (Crippa et al. 2004) as blood flow increases when brain regions are activated. CBD reduced the anxiety that participants reported during the neuroimaging. Consistent with the antianxiety effect of other known anxiety medications, CBD was found to modulate resting cerebral blood flow predominantly in limbic and paralimbic cortical areas, regions that are implicated in anxiety. A follow-up study with treatment-naive patients with social anxiety disorder found similar activity in limbic and paralimbic brain areas after CBD (400 mg, oral capsule) administration (Crippa et al. 2011). These results suggest that CBD's action in the brain is modulating brain activity in a similar manner as other antianxiety treatments and is doing so in known anxiety-related brain regions.

Previous neuroimaging studies in humans suggest that specific portions of the limbic system—the anterior cingulate cortex and the amygdala—are activated together in response to fear and anxiety. A double-blind, randomized, placebo-controlled study in fifteen healthy volunteers investigated the effect of CBD while participants viewed faces (which elicited different levels of anxiety) during functional magnetic resonance imaging (fMRI) scans. CBD (600 mg, oral capsule) reduced this connectivity between the anterior cingulate cortex and the amygdala while participants viewed fearful faces (Fusar-Poli et al. 2010), suggesting that changes in limbic system connectivity could be important in determining CBD's effects. However, further studies are needed to understand how CBD is exerting this effect in the brain and how this may be dependent on CBD dose, as well as the stimuli used to evoke anxiety.

## Conclusion

There is limited evidence to suggest that CBD may be beneficial in improving anxiety, but this evidence has methodological challenges.

Anecdotal reports and surveys suggest that CBD is indeed being used by the general public to manage their anxiety symptoms. Clinical trials that are currently underway aim to expand the dose range and evaluate whether chronic daily treatment with CBD can maintain its effectiveness in reducing anxiety symptoms. Further research is critical so that anxiety disorder patients and clinicians can make informed decisions regarding the use of CBD. Clinicians (and patients) also need to be mindful of the expectancy effect and how this affects the perceived efficacy of CBD in modulating anxiety.

# 9
# PTSD, Depression, and Sleep

Posttraumatic stress disorder, depression, and sleep are intimately connected. After experiencing a traumatic event, some individuals develop posttraumatic stress disorder (PTSD). Of those who do, approximately half also suffer from major depressive disorder (MDD). In addition, up to 90 percent of individuals with PTSD also report sleep disturbances such as nightmares. Multitarget treatments that can alleviate this collection of symptoms can be beneficial for these patients, and CBD may offer this therapeutic advantage. In this chapter, we discuss the evidence for CBD's effects in PTSD, depression, and sleep.

## CBD and PTSD

Soldiers, abused children, and women experiencing domestic violence are often affected by PTSD (Goldstein et al. 2016). The disorder is characterized by symptoms such as sleep disturbances, repeated recall of the event, mood disturbances such as depression and anxiety, as well as reduced social interaction. It is thought that the experience of trauma produces and consolidates traumatic memories through the association of environmental and interoceptive cues with the negative consequences of experiencing the trauma. Thus, PTSD is regarded as a disorder of learning and memory processes (Bowers and Ressler 2015; Ross et al. 2017). PTSD patients are more likely to consolidate or recall emotional or aversive memories, and they are impaired in their ability to properly extinguish associations between environmental cues and the negative aspects of traumatic stress exposure (see Careaga, Girardi, and Suchecki 2016 for review). PTSD often coexists with anxiety disorders, depression, sleep disturbances, and substance use disorders, further complicating treatment.

**CBD and PTSD: Preclinical Effects**

Currently there is no single accepted animal model of PTSD, but the fear-conditioning paradigm in rodents has been used to understand the processes of memory formation, which includes the stages of acquisition, consolidation, retrieval, reconsolidation, and extinction (LeDoux 2000; Maren and Quirk 2004). In the fear-conditioning paradigm, a neutral stimulus (the conditioned stimulus) such as a tone is paired with an aversive stimulus (the unconditioned stimulus) such as mild foot shock. After this pairing is made, the rodent learns that the conditioned stimulus (the tone) leads to the unconditioned stimulus (the foot shock). After one or more pairings of the conditioned stimulus with the unconditioned stimulus (for example, pairings of the tone with the foot shock), this association then leads to behavioral (for example, freezing) and physiological (for example, cardiovascular) responses. Due to this association, simply presenting the conditioned stimulus (the tone) will elicit the behavioral response (the conditioned response) of freezing. Fear extinction is the gradual attenuation of a previously learned association by repeated presentations of the conditioned stimulus without the unconditioned stimulus. This then creates a new memory that suppresses the original fear response. This process mimics that of exposure therapy used by clinicians, in which patients repeatedly experience a feared situation or stimulus (the conditioned stimulus) and learn that their feared consequence (the unconditioned stimulus) is unlikely to occur, eventually reducing fear. The use of CBD to interfere with various stages in the memory formation process has been examined using this model and is described below.

CBD treatment or infusion of CBD directly into brain areas of the fear circuit, before or after conditioning, reduces fear memory encoding (Levin et al. 2012; Norris et al. 2016; Stern et al. 2017; Rossignoli et al. 2017; Raymundi et al. 2020). That is, CBD interferes with the construction of fear memories. Unfortunately, blocking the creation of fear memories may not be clinically relevant or desirable as a treatment for PTSD, but it could be considered as a prophylactic treatment for special populations such as veterans entering combat.

Reducing the fear response, also known as attenuated learned fear expression, is demonstrated in rodents by decreased levels of freezing behavior. CBD treatment reduces learned fear expression (Resstel et al. 2006; Lemos, Resstel, and Guimaraes 2010; Jurkus et al. 2016; Song et al. 2016). When delivered to brain areas of the fear circuit, CBD also attenuates learned fear expression (Lemos, Resstel, and Guimaraes 2010; Gomes, Resstel, and Guimaraes 2011; Gomes et al. 2012; Fogaca et al. 2014), an effect that seems to be mediated by $5\text{-HT}_{1A}$ receptors (Gomes et al. 2012).

Once memories are formed and transformed into long-term memory (a process termed memory consolidation), these memories can be recalled and actively reformed by a process called reconsolidation. Reconsolidation serves to maintain, strengthen, and alter memories that are already stored in long-term memory. The process of reconsolidation can be disrupted by CBD treatment after memory retrieval (Stern et al. 2012; Stern et al. 2015; Murkar et al. 2019).

Finally, CBD treatment and direct infusion of CBD into brain regions of the stress circuit, enhance fear extinction (Do Monte et al. 2013; Bitencourt, Pamplona, and Takahashi 2008; Song et al. 2016), although one group has recently shown that CBD does not enhance the extinction of conditioned fear when using a predator-threat animal model of PTSD (Shallcross et al. 2019). These conflicting findings regarding CBD's effect on fear extinction should be further examined. In addition, only one study has examined the effect of CBD (10 mg/kg, i.p.) on fear learning in female mice, in comparison to the common PTSD SSRI treatment citalopram (Montoya, Uhernik, and Smith 2020). Like citalopram, CBD did not interfere with the formation or recall of the fear memory, but it did enhance fear extinction in female mice.

Taken together, preclinical evidence suggests that CBD interferes with various stages in the formation of traumatic memories. CBD reduces the formation, expression, and reconsolidation of these fear memories and may also facilitate the extinction of these fear memories.

**CBD and PTSD: Human Studies**

**CBD and Fear Conditioning in Healthy Volunteers** In a double-blind, placebo-controlled trial using a fear-conditioning paradigm (brief electric shocks as the unconditioned stimulus), Das et al. (2013) exposed forty-eight healthy human subjects to CBD (32 mg, inhaled). CBD enhanced fear extinction and attenuated fear responding. Although these data suggest that CBD could modulate multiple stages of aversive memory processing, the findings are in a relatively small number of healthy subjects and do not demonstrate clinical utility.

**CBD and PTSD Patients** A common limitation of these human studies is the small sample size used. Case reports follow only one patient, creating a significant sampling size bias. Recruitment of PTSD patients without comorbidities such as substance use disorder can often be difficult, as up to 40 percent of PTSD patients also have substance use disorder (Jacobsen, Southwick, and Kosten 2001). As these patients have been reported to have poorer treatment outcomes (Bowe and

Rosenheck 2015), they are an important group to include in such studies.

A randomized, double-blind, placebo-controlled, cross-over study examined the efficacy of three differing smoked cannabis varieties: high THC (12 percent THC, less than 0.05 percent CBD), high CBD (11 percent CBD, 0.5 percent THC), THC+CBD (7.9 percent THC, 8.1 percent CBD) compared to placebo in eighty military veterans for the treatment of PTSD after three weeks of treatment (Bonn-Miller et al. 2021). All treatment groups, including the placebo group, showed an improvement in PTSD severity, with no differences between placebo and other groups. These results highlight the need for further studies, given the strong placebo effect.

To date, there are only a handful of published reports of CBD use in PTSD patients. The first is a case report of a ten-year-old PTSD patient using CBD (25 mg oil and 6–12 mg spray, daily) to treat anxiety and sleep disturbances for five months (Shannon and Opila-Lehman 2016). CBD reduced anxiety and improved both the quality and quantity of the patient's sleep. In line with these findings, a retrospective case series examined the effect of CBD (25–100 mg capsule and/or 1–16 mg spray, daily for eight weeks, as an adjunct to the routine pharmacological and psychotherapy treatments) on PTSD symptoms in eleven adult patients (Elms et al. 2019). Ninety-one percent of patients experienced an improvement in their PTSD symptoms after eight weeks of treatment, and a subset of patients also reported a reduction in nightmares, a common symptom of PTSD. Because CBD was used as an adjunct treatment, we cannot assess the impact that these other pharmacotherapies had on the treatment outcome for these patients.

Finally, a fifteen-year-old survivor of acute sexual violence received CBD (300 mg, oral capsule, seven days) immediately after experiencing the trauma (Bolsoni et al. 2019). After a single dose of CBD, the patient still showed an increase in anxiety measures when recollecting the traumatic event. But after seven days of daily CBD treatment, the patient no longer showed an increase in anxiety scores when recollecting the traumatic event. This finding suggests that CBD may have interfered with the reconsolidation of the traumatic memory. It is important to note that CBD treatment was not able to prevent the development of PTSD in this individual but did seem to reduce the long-term symptoms of it. However, these studies were not randomized, double-blind, placebo-controlled trials, so there are limitations regarding biased results. Until additional double-blind, placebo-controlled, clinical trials are conducted with PTSD patients, we cannot determine CBD's effects on PTSD patients.

A randomized, double-blind, placebo-controlled clinical trial (www.clinicaltrials.gov, NCT03518801) is underway evaluating the use of CBD for sixteen weeks as an adjunct to prolonged exposure therapy (a type of psychotherapy that relies on extinction learning) in 136 US military veterans with PTSD at the VA San Diego Medical Center. At this time, no information on dosing is available, but follow-up will occur up to three months later. In addition, a randomized, double-blind, placebo-controlled clinical trial (NCT04197102) is also underway evaluating the use of CBD (300 mg oil, daily for eight weeks) alone or as an adjunct to trauma-focused group cognitive-behavioral therapy in 120 PTSD patients, with follow-up occurring up to three months. Findings from these two clinical trials should help to establish an optimal dose and dosing schedule for using CBD as a PTSD adjunct treatment.

**CBD and PTSD: Conclusion**
Preclinical studies suggest that CBD may interfere with the formation, expression, and reconsolidation of fear memories and may also facilitate fear memory extinction. The few human studies that have examined the effects of CBD in PTSD patients do not provide high-quality evidence. Ongoing, properly controlled clinical trials will help to elucidate CBD's effect in PTSD patients.

## CBD and Depression

Major depressive disorder (MDD) is a mood disorder characterized by depressed mood or anhedonia (the inability to find pleasure from normally enjoyable experiences) for at least two weeks. It is accompanied by a dysregulation in sleep and eating, psychomotor changes, feelings of guilt, and, in some patients, suicidal thoughts. MDD affects millions of people worldwide; almost half of patients diagnosed with MDD also have an anxiety disorder diagnosis, and many report the symptom of anxiety associated with their depressive episodes (Kessler et al. 2015).

Current antidepressant treatments can take several weeks to begin improving mood and therefore are limited in their efficacy (Cipriani et al. 2018). Research on new and more efficient drugs to treat depression is ongoing. Preclinical evidence suggests that CBD may rapidly produce antidepressant-like effects that can be sustained over time, perhaps making it a desirable therapeutic candidate or adjunct to evaluate in the clinic. Survey data indicate that depression is a medicinal use identified by CBD users (Corroon and Phillips 2018; Tran and Kavuluru 2020).

### CBD and Depression: Preclinical Studies

Depression is a multifaceted disorder with psychological, behavioral, and physiological symptoms, making it difficult to model in rodents. Therefore, behavioral tests that can reliably detect activity of known clinically effective antidepressants can be used to screen new potential treatments. The forced swim test (FST) and the tail suspension test (TST) are the experimental paradigms most often used in preclinical depression research, and both are based on the fact that after initial attempts to escape a stressful situation, the animal will stop such escape attempts and become immobile, indicative of learned helplessness. In the FST (Porsolt, Le Pichon, and Jalfre

1977), rodents are placed in a cylinder filled with water from which they cannot escape. Initially, the animals actively swim but then become immobile, and typical antidepressants reduce this immobility (Porsolt et al. 1978). In the TST (Steru et al. 1985) the rodent is hung by its tail, and after initial attempts to escape, the animal will then stop such escape attempts and become immobile; typical antidepressants reduce this immobility.

Using the FST, Zanelati et al. (2010) were the first to evaluate the antidepressant properties of CBD. They found that CBD had dose-dependent, antidepressant-like effects, with 30 mg/kg (i.p.) effectively reducing immobility time in mice. Similarly, El-Alfy et al (2010) and Florensa-Zanuy et al (2021) showed that CBD (30, 200 mg/kg, i.p.) had antidepressant-like effects in mice in the FST and/or TST, with CBD (30 mg/kg) also exhibiting pro-hedonic effects. Fluorinated CBD derivatives seem to have more potent antidepressant-like effects in FST: doses as low as 3 mg/kg (i.p.) were effective in mice (Breuer et al. 2016). Importantly, the antidepressant-like effect of CBD (3, 30 mg/kg, i.p.) seems to be maintained when administered for at least fourteen days (Schiavon et al. 2016; Reus et al. 2011; Sales et al. 2019), suggesting that chronic treatment with CBD may be effective in reducing depression-like effects in rodents without the development of tolerance. Finally, the epigenetic changes (such as DNA methylation) seen in brain regions relevant to depression (see Melas et al. 2021 for a review) can be reversed with CBD treatment in mice (Sales, Guimarães, and Joca 2020).

Research suggests that a dysfunctional serotonergic system may underlie major depressive disorder. CBD's antidepressant-like effects in rodents can be blocked by administering a $5\text{-HT}_{1A}$ receptor antagonist, suggesting that activation of the $5\text{-HT}_{1A}$ receptor mediates CBD's antidepressant-like effect (Zanelati et al. 2010; Chaves et al. 2021). Indeed, coadministration of an ineffective dose of CBD (7 mg/kg, i.p.) enhanced the effects of an ineffective doses of the classic antidepressant fluoxetine in mice in the FST (Sales et al. 2018).

This group also showed that inhibition of serotonin synthesis (reducing levels of serotonin) abolished CBD's effects in the FST, suggesting that CBD's antidepressant-like effects (at least in the FST) seem to depend on serotonin levels in the central nervous system. CBD (100 mg/kg, for seven days) treatment resulted in increased serotonin levels in the hippocampus (Abame et al. 2021). These finding suggest that CBD could be evaluated as an adjunct treatment with current antidepressants to restore levels of serotonin.

Studies using other animal models of depression are in agreement with the antidepressant-like properties of CBD. One such model, the learned helplessness model, uses exposure to inescapable shocks to induce poor escape performance, which can be mitigated by treatment with antidepressants. Indeed, CBD (30 mg/kg, i.p.) exerted antidepressant-like effects in rats using this model (Sales et al. 2019). In addition, using genetically modified mice (the olfactory bulbectomy model) and genetic rat models based on selective breeding (the Flinders Sensitive Line and the Wistar-Kyoto), CBD also showed antidepressant-like or pro-hedonic effects, or both (Linge et al. 2016; Shoval et al. 2016; Sales et al. 2019; Shbiro et al. 2019). Finally, HU-580, the methylated version of CBDA, also produced antidepression-like effects in the Flinders Sensitive Line and Wistar-Kyoto rats (Hen-Shoval et al. 2018). Together, these antidepressant-like effects of CBD in a number of preclinical models, using distinct rodent species and strains are promising findings that may guide future clinical trials.

## CBD and Depression: Human Studies

Currently there are no published properly controlled clinical trials investigating the effect of CBD on depressed patients. Despite the lack of empirical data, a recently published online survey suggests that approximately 15 percent of those using CBD reported using it for depressive symptoms (Corroon and Phillips 2018). This suggests that CBD is being used for depression management, although these are not necessarily clinically diagnosed patients.

A case study (Laczkovics et al. 2020) of a sixteen-year-old male with severe depression, social phobia, narcissistic personality disorder, and multiple substance use disorder (cannabis, cocaine, ecstacy) who was unresponsive to antidepressants showed improved depression and anxiety symptoms with CBD (100 mg up to 600 mg, oral capsules, over eight weeks) treatment. These improvements were seen with cessation of antidepressant medication.

The only other published study of CBD and depression (Solowij et al. 2018) involved a community sample of twenty regular cannabis users who were treated with CBD (200 mg, oral capsules, ten weeks) to investigate if it could restore psychological symptoms (as assessed by improvements from baseline measures versus week 10). Participants reported significantly fewer depressive and psychotic-like symptoms after ten weeks of CBD treatment. To understand these findings, properly controlled RCTs with larger samples are required. A double-blind, randomized, cross-over study (NCT04732169) is registered to evaluate the efficacy of Epidiolex in ten adults with major depressive disorder, over sixteen weeks.

**CBD and Depression: Conclusion**

A substantial number of preclinical studies suggest that CBD may have antidepressant-like effects in these rodent models. Currently only case studies have been published, and no properly controlled clinical trials have been published examining the effects of CBD in depression. One such registered trial is currently planned.

**CBD and Sleep**

Sleep plays an important role in our physical and mental health. It allows us to relax and restore our bodies and minds, and it is also important for the consolidation of memories into long-term storage. The process of sleep involves both nonrapid eye movement (NREM), or quiet sleep, and rapid eye movement (REM), or active sleep, both of which are important phases of sleep. Insomnia—the inability to obtain sleep that is sufficient to result in feeling rested or restored—is the most prevalent sleep complaint in the general population. Lack of sleep can lead to negative health consequences such as mood swings, depression, anxiety, and even cardiovascular disease.

Important aspects to those studying sleep and sleep disorders are sleep latency and REM sleep latency. Sleep latency—the amount of time from when the lights go out until we actually fall asleep—may be the most important parameter in assessing sleep. REM sleep latency is the time from sleep onset to the first instance of REM sleep and is therefore dependent on sleep latency. Alterations in REM sleep latency may be markers for sleep disorders.

Cannabis users commonly report noticing an improvement in sleep (Walsh et al. 2013). Twenty-five percent of recreational users reported using cannabis to help them relax and sleep (Lee, Neighbors, and Woods 2007). Indeed, in chronic pain patients aged fifty years and older, cannabis has been reported to have an overall positive effect on maintaining sleep at night (Sznitman et al. 2020). To parse out whether these self-reported positive sleep effects of whole cannabis are driven by THC or CBD, or both, may be therapeutically important. THC has been shown to reduce sleep latency (Cousens and DiMascio 1973). In contrast, those using cannabis strains with higher CBD concentrations (and therefore lower THC concentrations) reported insomnia and greater sleep latency (Belendiuk et al. 2015). Additionally, in this study, the need for sleep medication was more common for those using strains with lower THC concentrations (Belendiuk et al. 2015). Finally, a comparative study in palliative cancer patients showed improved sleep duration for those using higher THC cannabis strains (over higher CBD stains or mixed strains; Aviram et al. 2020). Taken together, these results suggest that in general, CBD may promote wakefulness with longer sleep latency, while THC may promote sleep with shorter sleep latency.

This may be why some medical cannabis patients tend to use CBD (or higher CBD strains) during the day and THC (or higher THC strain) at night (Piper 2018).

**CBD and Sleep: Preclinical Studies**

Preclinical studies suggest that CBD modulates sleep, but these effects seem to be dose dependent (see Murillo-Rodriguez et al. 2014 for review). Monti (1977) was the first to investigate the effects of CBD in rodents. In this early study, a moderate dose of CBD (20 mg/kg, i.p.) decreased sleep latency (with no effect on sleep or wakefulness time), while a high dose (40 mg/kg, i.p.) increased sleep time (and decreased wakefulness time) and decreased sleep latency. Neither dose of CBD altered REM sleep time or latency. Tolerance to these effects of CBD on sleep developed after fifteen days of treatment. More recent work has shown that rats treated with CBD (10, 40 mg/kg, i.p.) have an increase in their total percentage of sleep time, with a low dose of 2.5 mg/kg (i.p.) having no effect (Chagas et al. 2013). Furthermore, a moderate dose of CBD (10 mg/kg, i.p.) decreased REM sleep latency, while high-dose CBD (40 mg/kg, i.p.) increased it (Chagas et al. 2013). A low dose of CBD (2.5 mg/kg, i.p.) did not modify REM sleep latency (Chagas et al. 2013). In addition, the potent methyl ester version of CBDA, HU-580 (1.0 or 100 µg/kg, i.p.) prolonged wakefulness but did not modify REM sleep in rodents (Murillo-Rodriguez et al. 2020). These dose-dependent effects of systemic CBD administration on sleep and REM sleep in rodents generally suggest that high doses of CBD promote sleep, while low doses promote wakefulness. Certainly further research is needed to determine optimal doses for potential therapeutic use.

To shed further light on these seemingly contradictory effects of CBD when administered systemically, Murillo-Rodriguez et al. (2006) infused CBD directly into the cerebrospinal fluid in the cerebral ventricles by intracerebroventricular injection (i.c.v. infusion) in order to bypass the blood-brain barrier and distribute CBD throughout the brain. When administered i.c.v., CBD (10 µg/5 µL) increased wakefulness and decreased REM sleep. Furthermore, these wake-inducing effects seemed to be due to activation of neurons in the hypothalamus and the dorsal raphe nucleus, brain regions known to be involved in the generation of alertness (Murillo-Rodriguez et al. 2006, 2008).

Although the mechanism of action for CBD's wakefulness effects is still being investigated, evidence suggests an involvement of the dopamine system, as CBD (10 µg/1 or 5 µL, i.c.v.) in rats promotes an enhancement in the extracellular levels of dopamine in the nucleus accumbens, a brain region receiving direct projections from the hypothalamus to modulate sleep (Murillo-Rodriguez et al. 2006, 2011). Systemic injections of CBD (0, 5, 10, or 30 mg/kg, i.p.) promote an enhancement in the extracellular levels of acetylcholine in the forebrain, a brain region related to wake control (Murillo-Rodriguez et al. 2018). Furthermore, systemic injection of HU-580 (1, 100 µg/kg, i.p.) promotes an enhancement in the extracellular levels of dopamine and serotonin in the nucleus accumbens and an enhancement in the extracellular levels of acetylcholine in the forebrain (Murillo-Rodriguez et al. 2020). In summary, CBD (and HU-580) seem to promote wakefulness in rodents, likely through enhancement of extracellular levels of wake-related neurochemicals.

The effect of chronic adolescent exposure of CBD on sleep has been investigated. Treatment with CBD (5 or 30 mg/kg, i.p.) in adolescent rats for fourteen days resulted in increased adulthood wakefulness and decreased rapid eye movement sleep during the lights-on period (Murillo-Rodríguez et al. 2021). During the lights-off period, wakefulness was reduced, and slow wave sleep was increased. These animals also showed alterations in the sleep rebound period following total sleep deprivation. Indeed, further experiments are needed to clarify the potential mechanism of action of CBD on sleep modulation. It is possible that the dose-dependent effects of CBD on sleep are due to different mechanisms of action at low to moderate versus high doses in rodents. In addition, studies still need to examine whether the effects of CBD on sleep are maintained with chronic treatment in these preclinical models and further studies investigating early exposure to CBD are also important.

### CBD and Sleep: Human Studies

**CBD and Sleep in Healthy Volunteers** Converging evidence from a handful of clinical trials suggests there may be dose-dependent effects of CBD on sleep in healthy volunteers. That is, high doses of CBD (600 mg, oral capsule) induced sedative effects (Zuardi et al., 1993) or had no effect (300 mg, oral capsule) on sleep (Linares et al. 2018), while low doses (15 mg, oromucosal spray) increased wakefulness or had no effect (5 mg, oromucosal spray) on sleep (Nicholson et al. 2004). This suggests that CBD may have an alerting effect that is dose dependent. The lack of a CBD effect on sleep at some doses in healthy participants may suggest that CBD's beneficial effects may be realized only in those with dysregulated sleep cycles.

Indeed, in those with sleep disturbances such as difficulty in falling asleep and poor sleep throughout the night, a moderate dose of CBD (160 mg, oral capsule) increased total sleep time and decreased sleep interruption during the night (Carlini and Cunha 1981). Finally, in chronic pain patients, a low dose of CBD (30 mg daily for eight weeks, oral capsule) improved sleep quality (Capano, Weaver, and Burkman 2020), although it remains unclear whether this is due to an improvement in sleep per se or an improvement in the underlying pain condition.

**CBD and Sleep Disorder Patients** Only one study has reported the effect of CBD on sleep disorders in general. A retrospective case series (Shannon et al. 2019) was conducted at a psychiatric clinic involving clinical application of CBD (25–175 mg daily for three months, oral capsule) as an adjunct treatment for twenty-five patients diagnosed with any sleep disorder. In the first month of treatment, CBD improved sleep in 67 percent of patients, and at the three-month follow-up, sleep remained improved.

Narcolepsy patients experience excessive daytime sleepiness and cataplexy (sudden and involuntary muscle weakness or paralysis). Treatments for managing narcolepsy are compounds that induce alertness, suggesting that low doses of CBD may be a potential treatment for these patients. Using a rodent model of narcolepsy, CBD (5 mg/kg, i.p.) effectively blocked the sleepiness typically exhibited in these animals (Murillo-Rodriguez et al. 2019). No published clinical trials have evaluated the use of CBD for narcolepsy.

**CBD and PTSD Sleep Disturbances** PTSD patients commonly report increased arousal, difficulty sleeping, nightmares, and disrupted REM sleep. These patients often exhibit REM sleep abnormalities. To simulate the alterations to REM sleep shown by PTSD patients in rodents, Hsiao et al. (2012) used repeated anxiety testing in rats to produce anxiety-induced sleep disturbances and found that infusion of CBD (0.5, 1 µg) into the amygdala blocked the anxiety-induced suppression of REM sleep (Hsiao et al. 2012). These results are in line with a case study (Shannon and Opila-Lehman 2016) in PTSD patients showing that CBD (25 mg oil and 6–12 mg spray, daily for five months) reduced insomnia and sleep disturbances.

**CBD and Parksinon's Disease Sleep Disturbances** REM sleep behavior disorder is a common feature of Parkinson's disease, characterized by dream enactment during sleep. During sleep, patients physically move their limbs and may engage in activities such as talking, hitting, or punching. Chagas and colleagues (2014) investigated the effect of CBD (75–300 mg daily for six weeks) in reducing symptoms of REM sleep behavior disorder among four adults with Parkinson's disease. CBD successfully reduced their REM sleep behaviors, completely eliminating them in three of the four patients. A phase 2/3, double-blind, placebo-controlled, clinical trial investigated the effect of CBD (75–300 mg) in thirty-three patients with REM sleep behavior disorder for twelve weeks (de Almeida et al. 2021). No reduction in the frequency of REM sleep behaviors disorder symptoms occurred, but there was a significant improvement in average sleep satisfaction from the fourth to eighth week for CBD, in comparison to placebo. Limitations in these studies such as the single-center setting, the small sample sizes, and no serum CBD levels make these findings difficult to interpret, highlighting the need for further properly controlled clinical trials for sleep disturbances in Parkinson's disease patients.

**CBD and Sleep: Conclusion**
The field of CBD and sleep is in its early stages. To date, preclinical and early human work suggest biphasic effects of CBD on sleep. That is, low doses of CBD may have stimulatory effects to promote wakefulness, but high doses of CBD may have inhibitory effects to promote sleep. These findings suggest that the wake-promoting effects of low doses of CBD may be beneficial in narcolepsy (excessive daytime sleepiness). The few human studies that have been completed should be interpreted with caution as these studies are limited by small numbers of participants, short-term treatment and follow-up periods, and a lack of proper placebo controls, highlighting the need for further research into the effects of CBD on sleep and sleep disorders. Because of the possible dose-dependent effects of CBD on sleep, optimizing the doses for the desired effect (wakefulness versus sleep promotion) will be an important consideration.

## Conclusion

Some preclinical and clinical evidence suggests that CBD may be a potential therapeutic agent to investigate for improving symptoms of PTSD and depression and promoting wakefulness. Despite the lack of data from randomized clinical trials, anecdotal reports suggest that CBD is indeed being used by the general public to manage these conditions. Properly controlled human clinical trials, some currently underway, are needed to investigate CBD's effects more carefully in these conditions to potentially establish therapeutic dose ranges and evaluate whether chronic daily treatment with CBD could maintain its effectiveness. Further research is critical so that patients and clinicians can make informed decisions regarding the use of CBD as a treatment (or adjunct treatment) for these conditions.

# 10
# Psychosis and Schizophrenia

Psychosis is a symptom of numerous mental conditions and is not a specific illness. A psychotic patient exhibits loss of contact with reality and has delusions (false beliefs) and hallucinations (seeing or hearing nonexistent things). Speech and behavior are at times completely inappropriate. Psychosis is seen in schizophrenia and in the manic phase of bipolar disorder, but it may also occur in paranoia, postpartum psychosis, and various personality disorders. The use of drugs such as cannabis can also cause acute psychotic reactions. The *Diagnostic and Statistical Manual of Mental Disorders, Fifth Edition (DSM-5)* of the American Psychiatric Association emphasizes that these conditions occur along a spectrum, with schizoid personality disorder at the mild end and schizophrenia at the severe end. Psychosis is also identified as only one of several dimensions of neuropsychiatric disturbance in these disorders. Other dimensions encompass abnormal psychomotor behaviors, negative symptoms, cognitive impairments, and emotional disturbances (Arciniegas 2015). The symptoms are categorized as positive and negative. Positive symptoms include exaggerated ideas, perceptions, and actions that the person often cannot tell are real or not. The word *positive* indicates the presence (rather than absence) of symptoms that may include hallucinations (usually auditory), delusions, confused thoughts and disorganized speech, or movement disorders. Negative symptoms refer to the absence of normal mental function including lack of pleasure (anhedonia), lack of affect in speech, flattening of emotions or apathy, or social withdrawal. Someone with schizophrenia also may show cognitive dysfunction, including impairment in working memory and attention. Schizophrenia changes how one thinks, feels, and acts.

The symptoms of schizophrenia usually start between ages 16 and 30, often with a gradual change in behavior (prodrome phase) before obvious symptoms start. A widely accepted neurodevelopmental hypothesis of schizophrenia is that it is caused by the interaction of genetic and environmental factors, mostly during early neurodevelopment. This interaction can affect brain function and behavior and may be associated with specific and general impairments of cognitive function, leading to the mental disorder later in life (Owen et al. 2011).

Unregulated levels of the neurotransmitter dopamine (DA) and the $D_2$ receptor in certain brain areas are considered a major cause of psychosis/schizophrenia (Howes et al. 2017). As discussed in the mechanism-of-activity section in this chapter, CBD may modulate the mesolimbic dopamine system to produce antipsychotic effects.

Intake of high doses of $\Delta^9$-tetrahydrocannabinol (THC, the major psychoactive constituent of cannabis), particularly by first-time users of marijuana, may cause psychotic symptoms, even leading to hospitalization. By contrast, CBD does not cause psychotic effects. (For a detailed discussion of THC/CBD interactions, see chapter 3.)

Several reviews on various aspects of the effects of cannabinoids, including CBD, in psychosis/schizophrenia have been published recently (Bhattacharyya et al. 2012; Kucerova et al. 2014; Renard et al. 2017; Davies and Bhattacharyya 2019; Ghabrash et al. 2020; Schoevers, Leweke, and Leweke 2020). Those that have focused on CBD suggest that although there is some evidence for CBD reducing positive symptoms (such as delusions, hallucinations, and disorganized speech), only recently has evidence suggested that it may improve cognition (Lewke et al, 2021). CBD has a less severe adverse side effect profile in comparison to other antipsychotics and does seem to be well tolerated in patients. We first review the literature on the effects of CBD in preclinical animal models and then human clinical trial evidence.

## Preclinical Studies

It is quite difficult to model complex human disorders in animals. Many of the symptoms used to establish diagnoses such as schizophrenia (for example, hallucinations and delusions) are difficult to approximate in animals. Therefore, the use of certain animal models of schizophrenia-like behavior are thought to mirror only some aspects or symptoms of this complex disorder. Gururajan and Malone (2016) have summarized and evaluated the considerable number of publications on preclinical research using mice or rats with CBD in schizophrenia-like behavior. The methods used were mostly psychoactive drug–induced disruption of prepulse inhibition (PPI) or social withdrawal or hyperlocomotion; a few were on haloperidol-induced catalepsy and others on dexamphetamine-induced hyperlocomotion and dopamine effects using dopamine receptor agonists. All of these animal models are standard ones used for the various symptoms (positive, negative, cognitive) associated with schizophrenia-like behaviors (Nestler and Hyman 2010) and are discussed below.

MK-801 is an N-methyl-D-aspartate (NMDA) glutamate receptor antagonist known to produce hyperactivity, deficits in PPI, and social withdrawal. These behaviors correlate well with some of the positive, cognitive, and negative symptoms of schizophrenia. PPI is a model of attentional deficits in rodents. Rodents show a startle response (measured on an activity platform) to a strong reflex-eliciting tone (pulse), called an acoustic startle response. In PPI, a weak tone (prepulse) inhibits the reaction of an organism to a subsequent strong reflex-eliciting tone (pulse). When PPI is normal, the corresponding one-time startle response is reduced by the prepulse. However, deficits in PPI (produced by some psychoactive drugs such as MK-801 and amphetamine) are seen as failure of the prepulse to inhibit the response to the strong pulse. These deficits manifest in the inability to filter out unnecessary information. Such deficits are noted in patients suffering from schizophrenia. CBD has been shown to restore such deficits in PPI produced by psychoactive drugs (Long, Malone, and Taylor 2006; Pedrazzi et al. 2015).

A typical report on the MK-801-based assay by Gomes et al. (2014) found that MK-801 administration in mice at the dose of 1 mg/kg for twenty-eight days impaired PPI responses. MK-801 treatment also increased FosB/ΔFosB expression in the medial prefrontal cortex. Chronic treatment with CBD (30 and 60 mg/kg) attenuated the PPI impairment. All the molecular changes were also attenuated by CBD, which by itself did not induce any effect. This lack of effect has been confirmed by Schleicher et al. (2019), who found that prolonged CBD treatment caused no detrimental effects on memory, motor performance, and anxiety in C57BL/6J mice.

Another animal model of schizophrenia-like behavior is based on the pharmacological stimulation of dopaminergic neurotransmission, which is partially hyperactive in schizophrenia patients. In animals, apomorphine (a dopamine receptor agonist) is used to stimulate dopaminergic neurotransmission and is associated with the behavior of stereotypy characterized by repetitive motor actions that have no functional importance. An early study from the laboratory of Antonio Zuardi (Zuardi, Rodrigues, and Cunha 1991) showed that in rodent models at a dose of 60 mg/kg, CBD was able to attenuate apomorphine-induced stereotypy.

Some antipsychotic drugs cause extrapyramidal side effects due to dopamine $D_2$ receptor blockade of the nigrostriatal pathway (Kapur and Remington 2001). These effects include dystonias and dyskinesias; in rodents, they can manifest as cataleptic behaviors, such that if an animal is placed with its forepaws resting on a suspended horizontal bar, it will remain frozen in that position for an extended period of time. The anticataleptic effect of CBD was studied in Swiss mice by Gomes et al. (2013). CBD (5–60 mg/kg) pretreatment was able to dose-dependently inhibit the cataleptic effects of the antipsychotic drug haloperidol (0.6 mg/kg), which is well known to produce such effects. Injection of CBD into the dorsal striatum inhibited haloperidol-induced catalepsy (Sonego et al. 2016). The anticataleptic effect was also inhibited by the $5\text{-HT}_{1A}$ receptor antagonist WAY100635, when it was administered as pretreatment to CBD, suggesting an involvement of the 5-HT system in CBD's anticataleptic effects. No side effects typical for antipsychotic drugs were noted with CBD.

Patients with schizophrenia at times have psychomotor agitation that manifests itself as aggression. A similar effect can be seen in mice by injections of psychostimulants such as dexamphetamine and ketamine, which induce hyperactivity. A high dose of CBD (60 mg/kg) has been reported to inhibit hyperlocomotor activity, induced by dexamphetamine or ketamine (Moreira and Guimaraes 2005). Intra-accumbal administration of CBD inhibits the disruption caused by amphetamine (Pedrazzi et al. 2015). Acute administration of haloperidol to rats increases gene c-fos expression in some brain regions, including the dorsal striatum, which is involved in hyperlocomotive effects. Injections of CBD into this brain region in mice attenuates the haloperidol-induced increases in c-fos expression (Sonego et al. 2016).

In agreement with the neurodevelopmental hypothesis of schizophrenia, prenatal exposure of rats to the antimitotic agent methylazoxymethanol acetate (MAM) at gestational day 17 produced long-lasting behavioral alterations such as social withdrawal and cognitive impairment in the social interaction test and in the novel object recognition test, respectively (Stark et al. 2019). At the molecular level, increased $CB_1$ receptor mRNA and protein expression coincided with deficits in the social interaction test and in an object recognition test in MAM rats. Both the schizophrenia-like phenotype and altered transcriptional regulation of $CB_1$ receptors were reversed by peripubertal treatment with CBD. The authors suggest that these results indicate that early treatment with CBD may prevent the appearance of schizophrenia-like deficits at adulthood.

Levin and colleagues (2014) have shown that a spontaneously hypertensive strain of rats (SHR) can serve as an animal model of schizophrenia. In these rats, a deficit of PPI is observed, which parallels the effect of drugs described above. Treatment with CBD attenuated the PPI impairment. The same group also reported that in SHR rats, CBD administered during periadolescence prevented the emergence of hyperlocomotor activity (a model for the positive symptoms of schizophrenia), as well as deficits in PPI of startle and contextual fear conditioning (cognitive impairments; Peres et al. 2018). In the SHR assay, CBD exhibited an anxiolytic but not antipsychotic property evaluated in the social interaction test (Almeida et al. 2013). The results with CBD in SHR rats also indicated the involvement of the serotoninergic system in these effects.

Taken together, these preclinical rodent models suggest that CBD treatment may reduce or restore some of these behaviors that approximate some of the symptoms seen in schizophrenia patients. However, caution must be exerted in generalizing the findings from these preclinical animal models to the complex mental human disorder called schizophrenia. These animal models will never fully mirror the complexities of this exclusively human disorder of the mind.

## Human Clinical Studies

On the basis of animal tests by Zuardi et al. (1991) summarized above, which showed that CBD may reduce some behaviors that approximate psychosis in humans, the same group (Zuardi et al.

1995) administered CBD to a nineteen-year-old schizophrenic patient. The dose was gradually increased over twenty-six days up to 1,500 mg/day. During periods of great agitation and/or anxiety, oral diazepam 10 mg was also administered. The effect of CBD was evaluated by three different psychiatric rating scales, all of which indicated significant clinical improvement, comparable to (or even better than) that noted with haloperidol. However, in a further trial (Zuardi et al. 2006) when CBD was administered as monotherapy to three treatment-resistant schizophrenia patients, slight improvement was seen in only one patient. The authors concluded that CBD may not be efficient in treatment-resistant schizophrenia patients. In a trial with CBD treatment in thirty-three patients with bipolar affective disorder, the same group (Zuardi et al. 2010) again found that CBD was ineffective.

In contrast, in a phase 2, double-blind, randomized, parallel group design, Leweke and colleagues (2012) compared the effect of CBD with that of amisulpride, a potent and widely used antipsychotic drug, in forty-two acutely exacerbated paranoid schizophrenia patients. The CBD group was administered initially 200 mg/day which, within a week, was gradually increased to 800 mg/day. This therapeutic dose was administered for an additional three weeks. The clinical measurements were based on two scales, the Positive and Negative Syndrome Scale (PANSS scale) and the Brief Psychiatric Rating Scale (BPRS scale). CBD was as effective as amisulpride in improving the psychotic symptoms and even displayed a markedly superior side effect profile. It had marked tolerability.

The same group had previously noted an elevation of anandamide levels in the cerebrospinal fluid of psychosis patients compared to normal individuals (Giuffrida et al. 2004). Anandamide is a major endogenous cannabinoid (Devane et al. 1992). The level of anandamide was inversely correlated with the psychotic symptoms. Indeed, the authors state that "enhanced anandamide signaling led to a lower transition rate from initial prodromal states into frank psychosis as well as postponed transition." Moreover, CBD treatment was accompanied by a significant increase in serum anandamide levels, which was significantly associated with clinical improvement (Leweke et al. 2012). The mechanism of the enhancement of anandamide is apparently due to the blocking effect by CBD of the hydrolysis of anandamide. Indeed, there was a significant negative correlation between serum anandamide levels and PANSS score in CBD-treated patients. The anandamide observations may indicate that the antipsychotic effects of CBD are actually due to anandamide. However, this mechanism should be interpreted with caution in light of a recent report that although CBD effectively inhibited rat FAAH, it was much less effective as an inhibitor of human FAAH (Criscuolo et al. 2020). Anandamide, an endogenous neurotransmitter, has not been investigated as an antipsychotic molecule in either healthy individuals or in patients.

Positive results in the treatment of schizophrenia patients with CBD were also published by McGuire et al. (2018). In a double-blind clinical trial, the effectiveness and safety of the CBD treatment were assessed. CBD (1 g/day for six weeks) was administered to forty-three patients; placebo was administered to forty-five patients. Both groups continued to receive their previously used antipsychotic medication. All patients were assessed using the PANSS scale, as well as four additional scales: the Global Assessment of Functioning scale (GAF), the Brief Assessment of Cognition in Schizophrenia (BACS), and the improvement and severity scales of the Clinical Global Impressions Scale (CGI-I and CGI-S). On the basis of these scales, the patients who received CBD were found to have an improved medical profile, with lower levels of positive psychotic symptoms (PANSS), as well as improvement noted in the CGI-I and CGI-S scales. Improvements that did not reach statistical significance were noted in cognitive performance (BACS) and in overall functioning (GAF). CBD was well tolerated, and the adverse events did not differ much between the CBD and placebo groups.

Bhattacharyya and colleagues (2018) have reported results of a double-blind study of the effect of CBD in thirty-three clinical high-risk (CHR) psychosis patients. CBD caused a significant reduction of distress. A trend toward reduction of the severity of psychotic symptoms was also noted. CBD partially normalized alterations in parahippocampal, striatal, and midbrain function associated with the CHR state. Because these regions are critical to the pathophysiology of psychosis, the authors conclude that CBD at these sites may underlie its therapeutic effects on psychotic symptoms.

The mechanism of these mostly positive results with CBD has not been established. McGuire et al. (2018) found that it did not depend on dopamine receptor antagonism, the receptor mediating many of the antipsychotic drug effects.

Contrary to the positive results of the clinical trials already summarized, negative results have also been published. Boggs et al. (2018) compared the cognitive, symptomatic, and side effects of CBD (600 mg/day) in thirty-six chronic and stable out-treated schizophrenia patients versus placebo-controlled patients over a six-week period. All patients continued to receive their previous antipsychotic drugs. Cognition was evaluated by the MATRICS Consensus Cognitive Battery (MCCB); the PANSS scale was used to assess the psychotic symptoms. No improvement in cognition was observed. No significant improvement in either the MCCB or PANSS scores was noted. No significant side effects were observed. It is possible that the differences between the negative data by Boggs et al. (2018) and the positive data by Leweke et al. (2012), McGuire et al. (2018), and Bhattacharyya et al. (2018) are due to either the lower doses administered by Boggs et al. or to interactions with other antipsychotic drugs administered. Indeed, Leweke et al. (2021) recently reported that a dose of CBD up to 800 mg/day improved neurocognitive performance in a parallel-group, active-controlled, monotherapeutic, double-blind, randomized-control trial in forty-two acute paranoid schizophrenic patients receiving either CBD or amisulpride for four weeks in an inpatient setting. Both drugs increased visual memory and processing speed. CBD improved sustained attention and visuomotor coordination, while amisulpride enhanced working memory performance. Both treatment drugs improved neurocognitive functioning in acutely ill schizophrenia patients. Therefore, it is conceivable that the prior negative findings (Boggs, Surti et al. 2018) may be related to the lower dose of CBD or to the add-on study design (maintaining the patients on their standard treatment) employed. However, the lack of a placebo control condition (Leweke et al. 2021) does not allow for an estimate of a potential placebo response to both CBD and amisulpride. Larger RCTs are needed to confirm the exploratory finding that CBD may improve cognitive performance in schizophrenic patients.

Hundal and colleagues (2018) have published that CBD (600 mg, single administration) did not lower anxiety or persecutory ideation in thirty-six healthy volunteers with high trait paranoia. Zuardi et al. (2010) also found it to be ineffective in two patients in the manic episode of bipolar affective disorder. No systematic studies of CBD in mood disorders (including bipolar disorder) have been undertaken, although some patients claim that cannabis relieves symptoms of mania or depression (Ashton et al. 2005; Pinto et al. 2020). The results of a few additional minor clinical trials with CBD are summarized by Davies and Bhattacharyya (2019).

There is considerable recent interest in the potential of CBD to serve as a treatment for schizophrenia as revealed by the myriad of studies currently being conducted that are listed on the NIH website www.clinicaltrials.gov. There are currently over fifty active clinical trials at various stages listed that are evaluating the effect of CBD on human psychosis. Over the next few years, we expect that these trials will provide a better understanding of the efficacy of CBD as a treatment, either alone or as an adjunct therapy, for this intractable mental disorder.

## Mechanisms of Action

Dysregulated levels of the neurotransmitter dopamine and the $D_2$ receptor in certain brain areas are considered a major cause of psychosis/schizophrenia. In the mesolimbic pathway, they may lead to the positive symptoms of schizophrenia; in the mesocortical pathway, they may be responsible for the negative symptoms of the disease.

Essentially all currently effective antipsychotic pharmacotherapies for schizophrenia target the brain dopamine receptor system directly or indirectly. There is also considerable evidence that the action of CBD on psychosis, specifically on schizophrenia, is due to a large extent to modulation of the mesolimbic dopamine (DA) system, with involvement of the 5-HT system. Renard and colleagues (2016) have shown that CBD counteracts amphetamine-induced neuronal and behavioral sensitization of the mesolimbic DA pathway through a novel kinase signaling pathway.

The clinical and preclinical evidence demonstrating the modulatory effects of CBD on DA activity, primarily within the mesolimbic pathway, has been reviewed (Renard et al. 2017). Functional interactions of CBD as an agonist with the serotonin 5-$HT_{1A}$ receptor system are likewise described. However, we still need considerably more thorough and detailed knowledge of the molecular and neuronal action of CBD on both the DA and 5-HT systems. Norris and colleagues (2016) have shown that CBD in the mesolimbic nucleus accumbens is capable of blocking the formation of associative fear memories in rats through a 5-$HT_{1A}$ receptor-dependent mechanism. However, CBD also modulates neuronal network dynamics directly in the ventral tegmental area (VTA) by decreasing the spontaneous frequency and bursting activity levels of the VTA dopaminergic neurons.

THC is known to induce schizophrenia-like deficits associated with attenuated activation in the cortical, striatal, and amygdala brain regions. Bhattacharyya et al. (2010) reported that CBD administration reversed these effects both behaviorally and in neural activation patterns. THC versus CBD produced opposite effects on schizophrenia-like attentional salience processing deficits in the hippocampus and caudate nucleus (Bhattacharyya et al. 2012). However, the behavioral changes caused by CBD in rodent models of schizophrenia have been shown recently not to be due to effects on the cannabinoid receptors but to effects on the 5-HT$_{1A}$ receptor (Rodrigues da Silva et al. 2020). As well, Hudson et al. (2019) have recently reported that the potential of CBD to reverse the effects of THC on psychotic-like responding in rats may be through bidirectional control of the molecular ERK1–2 pathway in the control of DA release by the VTA. Using in vivo electrophysiology, the authors found that THC enhanced such release, but CBD prevented the effect of THC on DA activity. O'Neill et al. (2020) have recently published data showing that normalization of mediotemporal and prefrontal activity, and mediotemporal-striatal connectivity, may underlie the antipsychotic effects of CBD.

CBD may cause its antipsychotic effects via upregulation of the endocannabinoid anandamide, presumably by inhibition of fatty acid amide hydrolase (FAAH), its degrading enzyme (Leweke et al. 2012); however, recent in vitro evidence suggests that CBD may not efficaciously inhibit human FAAH (Criscuolo et al. 2020) as it has been shown to effectively inhibit rat FAAH.

Some of the actions of CBD are through other biological systems, some of which may be relevant to its antipsychotic effects. For example, CBD stimulates vanilloid receptor type1 (VR1) (Bisogno et al. 2001). Surprisingly, its enantiomer (+) CBD has the same activity. CBD binds weakly to the cannabinoid CB$_1$ and CB$_2$ receptors (Pertwee 2008). CBD also binds to GPR55 (Ryberg et al. 2007), but there are no published indications that these effects are relevant to its antipsychosis/schizophrenia action.

## Conclusion

In a recent, outstanding review, Schoevers et al. (2020) summarized the action of CBD in psychosis/schizophrenia. Their summary parallels to a large extent our view:

A. There is a role for CBD in the treatment of psychosis/schizophrenia. Almost all clinical publications indicate that CBD affects numerous positive disease symptoms.
B. There is only limited recent evidence (Leweke et al. 2021) that cognition may be impacted
C. The clinical trials reported so far have lasted a limited number of weeks. We have no evidence whether tolerance develops to CBD. Long-term studies are needed.
D. CBD is well tolerated, even at high doses; however, it is unknown whether toxicity develops following long-term access to such high doses, especially in light of interaction with liver enzymes.

From a therapeutic point of view, the preclinical and clinical evidence showing that CBD may normalize affective deficits associated with schizophrenia models in rodents and positive effects on schizophrenia patients indicates that it may represent a promising treatment for schizophrenia, acting through novel molecular and neuronal mesolimbic pathways. Most recently, Chesney, Oliver and McGuire (2021) have reported that CBD may be a novel treatment in the early phase of psychosis. The pharmacological interventions available in this phase are extremely limited among individuals at high clinical risk for psychosis. CBD may be an ideal treatement for these prodromal individuals.

# 11
# Addiction

Drug addiction is a motivational disorder in which an individual compulsively seeks to use drugs and loses control over drug intake (Koob and Volkow 2010). It is a chronically relapsing disorder characterized by compulsion to seek and take the drug, loss of control in limiting drug intake, and the development of a negative emotional state (including anxiety, anhedonia, and irritability), which represents a motivational withdrawal state when access to the drug is prevented (Koob and Volkow 2010). One of the greatest challenges in the treatment of addiction is relapse to drug use after a period of withdrawal. Environmental cues associated with the effects of drugs are one of the strongest triggers for craving, which contributes to relapse. Among the classes of abused drugs, there are few effective treatments available. In preclinical studies in animals, evidence is growing that CBD may show promise to reduce the potential of these drug-associated environmental cues to trigger relapse (Parker et al. 2004; Ren et al. 2009; Hurd et al. 2019). Human clinical trial research, however, has produced conflicting results (Mongeau-Pérusse et al. 2021; Hurd et al. 2019).

## Preclinical Animal Models of Addiction

Before the testing of CBD as a treatment for addiction in humans, it was evaluated for its potential to modify patterns of drug taking, drug seeking, and relapse in animal models, generally using rats or mice. It is important to note that the same drugs that act as rewards in humans are also rewarding to other animals in these models. The two best-characterized rodent models of the rewarding effects of drugs are the self-administration procedure and the conditioned place preference paradigm. In the drug self-administration paradigm, an animal must learn to press a lever to deliver an intravenous infusion of the drug through an implanted catheter. If the drug is rewarding (for example, morphine, heroin, cocaine, amphetamine), the rodent will learn quickly to press the lever to deliver the drug. Once the rodent learns to deliver the drug, the motivation to self-administer the drug is assessed by requiring more effort (lever presses) to receive the reward. This is a measure of drug taking, or how hard the animal will work to get the drug. Through a process of classical conditioning, stimuli (such as a light) that are paired with the infusion of the drug acquire conditioned reinforcing properties as they signal upcoming access to the drug. These stimuli can subsequently be presented to the animal to elicit drug-seeking behavior, which means responding in anticipation of receiving the drug (craving). This paradigm is useful in evaluating medicines for controlling drug craving, which is believed to promote relapse to drug taking in humans. Relapse is often measured by extinguishing the response of lever pressing by lack of reinforcement (the lever is presented, but the drug isn't available) and then reinstating the response by presentation of the drug-paired conditioned cues (cue-induced reinstatement) or by injecting the drug prior to the session (drug-induced reinstatement).

The conditioned place preference paradigm is another model of drug reward using rodents. A rodent is injected with a drug and placed in a distinctive chamber on one day and injected with a placebo and placed in a different distinctive chamber on another day, with the order counterbalanced. This procedure is usually continued until the rat has received four pairings of the drug with the chamber. Then the animal, when drug free, is given a choice between the two chambers. The rodent will spend more time in the drug-paired chamber if the drug was rewarding but less time in that chamber if the drug was aversive.

CBD has been evaluated for its addictive potential in both the self-administration paradigm (Ren et al. 2009) and the conditioned place preference paradigm (Parker et al. 2004) and found neither to be self-administered nor to produce a place preference or aversion on its own. However, CBD has been shown to modify the rewarding effects of other abused drugs.

## Opiates

Opiate addiction has been a primary target for investigation of the therapeutic potential of CBD treatment. CBD reduces opiate withdrawal signs in opiate-dependent animals (Chesher and Jackson 1985; Hine, Friedman et al. 1975; Hine, Torrelio, and Gershon 1975; Bhargava 1976). Although CBD does not reduce heroin self-administration ("drug taking"), it reduces the potential of a heroin-paired cue to trigger "drug seeking" for weeks following CBD administration (Ren et al. 2009). CBD also normalizes heroin-induced impairments in the brain reward systems (specifically in the nucleus accumbens; Renard et al. 2016, 2017).

On the basis of the promising preclinical research, a series of clinical studies conducted in humans showed that CBD was safe in humans and did not result in adverse consequences when coadministered with the potent synthetic opiate agonist fentanyl, even at high doses (Manini et al. 2015; Taylor et al. 2018). In humans, CBD did not modify the subjective effects of fentanyl but reduced craving and anxiety triggered by heroin-associated cues (Hurd et al. 2015). Most recently, Hurd and colleagues (2019) used a double-blind, randomized, placebo-controlled trial to explore the effects of acute and short-term CBD administration on craving and anxiety in abstinent men and women with heroin use disorder. Craving was triggered by a video of heroin-associated environmental cues. CBD (400 mg or 800 mg oral Epidiolex) or placebo was administered sixty minutes before the first cue test and the test was repeated twenty-four hours later and one week later. Since CBD may reduce anxiety (see chapter 8), it was expected to reduce stress responsivity (determined by heart rate, respiratory rate, and salivary cortisol levels) and craving (determined by questionnaire) produced by the heroin-paired cues. Since it has also been shown that CBD reduces attentional bias to cigarette cues in tobacco smokers (Hindocha et al. 2018), it was also expected to reduce salience of the heroin-associated cues. Both doses of CBD reduced cue-induced craving, anxiety, and physiological measures of stress reactivity. The strongest effects were seen sixty minutes after administration, but even one week after the administration of CBD, the participants showed reduced craving and anxiety, suggesting that as in the preclinical results (Ren et al. 2009), the effects of CBD on craving may be long lasting. Indeed, a recent preclinical finding (Gonzalez-Cuevas et al. 2018) confirmed the prolonged effects of CBD on drug (alcohol and cocaine) seeking and anxiety-like behaviors five months after its short-term administration. This long-lasting property of CBD could have significant clinical implications, especially for patient populations in which daily medication adherence may be challenging. While these findings are encouraging, they do not represent an efficacy trial for opioid addiction and should not be used as evidence to forgo treatments that already exist. More research on the efficacy of CBD specifically in reducing opiate dependence over a longer term is severely needed.

In vivo neuroimaging has indicated that CBD blunts activity in limbic neural circuits engaged during negative emotional processing

(Fusar-Poli et al. 2009) and modulates networks linked with attention salience processing (Bhattacharyya et al. 2012). These findings suggest that CBD does not directly modify the rewarding effects of opiates, but may reduce the potential of cues associated with opiates to induce craving, a classic relapsing phenomenon.

## Nicotine

In the case of nicotine addiction, preliminary findings of a placebo-controlled study of twenty-four smokers (more than ten cigarettes a day), showed that those who received a CBD inhaler (400 μg/dose) reduced the number of cigarettes smoked relative to a placebo group, but there was no reduction of craving or withdrawal (Morgan et al. 2013). This may be related to the ability of CBD to inhibit FAAH, thereby elevating anandamide and other fatty acids (such as N-palmitoylethanolamide and N-oleoylethanolamide). In preclinical animal work, FAAH inhibitors have been shown to prevent nicotine seeking in rats (Forget 2009; Scherma 2008); however, recent in vitro findings suggest that CBD is more effective in inhibiting FAAH in rats than in humans (Criscuolo et al. 2020). It has also been found that oral CBD (800 mg) reduced the salience of cigarette cues after overnight abstinence in smokers relative to placebo, but they did not find a reduction in craving or withdrawal (Hindocha et al. 2018). Together these findings suggest that CBD may reduce nicotine seeking but may not be beneficial in reducing cravings or the symptoms of withdrawal.

## Alcohol

Preliminary preclinical results suggest that CBD can attenuate alcohol consumption and potentially protect against certain harmful effects of alcohol, such as liver and brain damage (Nona, Hendershot, and Le Foll 2019; De Ternay et al. 2019). Preclinical studies of CBD's effects on alcohol outcomes have evaluated alcohol consumption and motivation to use alcohol, relapse and withdrawal, and protection from alcohol-induced liver and brain damage. To evaluate the effect of CBD on oral self-administration of alcohol, mice were given a single subcutaneous injection of controlled release 20 mg/kg/day CBD prior to the start of a series of self-administration tasks. The CBD produced a significant reduction in alcohol consumption and the number of active lever presses for alcohol (Viudez-Martinez, Garcia-Gutierrez, Fraguas-Sanchez, et al. 2018). However, when the rats were required to work hard for the opportunity to obtain the alcohol in a more effortful task (progressive ratio schedule of reinforcement), it was no longer effective in reducing motivation to drink. On the other hand, a subsequent study by the same group (Viudez-Martinez, Garcia-Gutierrez, Navarron, et al. 2018) demonstrated that a very high dose of CBD (120 mg/kg) did attenuate motivation to work effortfully for alcohol. The mechanism for this effect was not investigated.

The attempt to remove unpleasant withdrawal symptoms is one reason for relapse to drug use. There is evidence that CBD may reduce alcohol withdrawal and relapse behaviors. Preclinical studies have shown that CBD (60 mg/kg; i.p.) is effective in reducing ethanol self-administration and at a very high dose (120 mg/kg; i.p.), attenuated ethanol relapse (Viudez-Martinez, Garcia-Gutierrez, Navarron et al. 2018). Indeed, seven daily administrations of CBD (15 mg/kg, transdermal) also reduced cue- and stress-induced relapse to ethanol administration up to 138 days post-CBD treatment (Gonzalez-Cuevas et al. 2018). CBD did not produce sedation or interrupt normal motivated behavior (observed by normal self-administration of sucrose solution).

CBD also reduced handling-induced seizures in mice undergoing ethanol withdrawal (Sanmartin and Detyniecki 2018). Alcohol-related liver damage includes hepatosteaosis, which can lead to hepatitis and cirrhosis, while alcohol-induced damage to the brain includes neuroinflammation, cell death, and cognitive deficits (Nona, Hendershot, and Le Foll 2019). CBD (5 mg/kg, i.p.) given thirty minutes prior to an alcohol binge model for five to eleven days has been shown to reduce hepatic injury in multiple markers of liver injury (Yang et al. 2014; Wang et al. 2017). As well, CBD has shown promise to reduce alcohol-induced neurotoxicity. A high dose of CBD (40 mg/kg, i.p.), but not a lower dose (20 mg/kg, i.p.), administered for three days of a four-day alcohol binge attenuated damage in the hippocampus and entorhinal cortex (Hamelink et al. 2005). As well, transdermal gel applications (5 percent and 2.5 percent CBD) or twice-daily 20 mg/kg, i.p. injections were also effective in reducing hepatic damage (Liput et al. 2013). CBD also stimulates autophagy, a mechanism that can protect liver cells from alcoholic liver disease (Yang et al. 2014).

The only human studies on the interaction of CBD with alcohol were conducted in the late 1970s and early 1980s (Belgrave et al. 1979; Consroe et al. 1979; Bird et al. 1980). These studies consisted of healthy social drinkers given a battery of tests under the influence of placebo, alcohol, CBD, and a combination of alcohol plus CBD. CBD alone had no effect on motor or psychomotor performance and produced no subjective effects. CBD (200 mg) given simultaneously with alcohol (Consroe et al. 1979) reduced blood alcohol levels. However, none of these studies indicated that CBD interfered with alcohol intoxication. There have been no high-quality human clinical trials on the potential of CBD to interfere with alcohol intake or alcohol-induced liver or brain damage. However, with the recognition of the therapeutic potential of CBD for opiate addiction, clinical trials in adults with severe alcohol use disorder (https://clinicaltrials.gov/ct2/show/NCT03252756) and co-occurring alcohol use disorder and posttraumatic stress disorder (https://clinicaltrials.gov/ct/show/NCT03248167) are ongoing. These trials may hold hope for the use of CBD in treating this disorder.

## Cannabis

Cannabis use disorder is characterized by withdrawal symptoms following cessation of use, including sleep disturbances, decreased appetite, irritability, increased anxiety, and depressed mood. In human case studies, CBD eliminated self-reported cannabis use in a dependent male (Shannon and Opila-Lehman 2015) and reduced cannabis withdrawal in another (Crippa et al. 2013). Finally, an open label clinical trial (Solowij et al. 2018) assessed the effects of ten-week treatment with CBD (200 mg/day) on psychological symptoms, cognition, and plasma concentrations among twenty frequent and ongoing cannabis users (twelve were diagnosed with cannabis use disorder). Cannabis use and withdrawal did not change between baseline and posttreatment sessions, but cannabis-related experiences (feeling high) decreased as a result of CBD treatment. Higher CBD plasma concentrations were associated with lower psychotic-like symptoms, distress, anxiety, and severity of cannabis dependence. These results suggest greater effects of CBD in dependent users, which can possibly be detected through CBD plasma concentrations. Therefore, CBD shows some promise in the treatment of cannabis dependence, but double-blind, placebo-controlled, randomized, control trials with pure CBD are lacking.

Most work on cannabis use disorder has focused on nabiximols, an oral mucosal spray containing 2.7 mg of THC and 2.5 mg of CBD. In a double-blind randomized, controlled trial, 51 inpatients with cannabis dependence received nabiximols or placebo for six days, along with cognitive behavioral therapy. Immediately after treatment, nabiximols decreased cannabis withdrawal and craving symptoms, but there was no difference among the groups in cannabis use twenty-eight days later (Allsop et al. 2014). A second double-blind, randomized, placebo-controlled trial assessed the effects of an eight-week treatment with self-titrated or fixed doses of nabiximols in nine participants with cannabis dependence (Trigo et al. 2016).

During the treatment sessions, both fixed and self-titrated doses of nabiximols reduced cannabis withdrawal symptoms; however, craving was not affected. A later, larger double-blind study by this same group treated twenty-seven cannabis-dependent participants with self-titrated dosages of nabiximols in combination with cognitive behavioral therapy over twelve weeks (Trigo et al. 2018). In this study, the abstinence rate did not change significantly between baseline and follow-up, but nabiximols was associated with a greater reduction in cannabis-craving symptoms when compared with placebo. These early studies suggested that nabiximols may be effective in reducing withdrawal but not improving abstinence rate (Trigo et al. 2018; Allsop et al. 2014; Trigo et al. 2016). However, with a long-term treatment of twelve weeks, a more recently published randomized clinical trial (Lintzeris et al. 2019) showed that the nabiximols group reported fewer days using cannabis over the treatment period than the placebo group. This study suggests that combined CBD/THC may reduce cannabis use in individuals with cannabis use disorder.

## Stimulants

In preclinical studies, the effects of CBD on cocaine (Galaj et al. 2020, Mahmud et al. 2017) and methamphetamine (Hay et al. 2018) self-administration appear to be dose dependent, with high doses (20–80 mg/kg, i.p.) or repeated treatments (Lujan et al. 2018, Gonzalez-Cuevas et al. 2018) needed to see suppression of both drug taking and drug seeking (Calpe-Lopez, Garcia-Pardo, and Aguilar 2019). As well, CBD prevented priming and stress-induced reinstatement of a CPP induced by cocaine in mice (Calpe-López et al. 2021). The only randomized placebo-controlled trial of CBD in treatment of craving and relapse (Mongeau-Pérusse et al. 2021) concluded that CBD (800 mg/day) for twelve weeks did not reduce cocaine craving or relapse among people being treated for cocaine use disorder. However, a recent randomized controlled trial (Morissette et al. 2021) revealed that CBD may exert anti-inflammatory effects in individuals with cocaine use disorder displaying oxidative stress–induced inflammatory response.

## Conclusion

The strongest human clinical evidence for the promise of CBD as a treatment for addiction has been provided for opiate addiction (Hurd et al. 2019), with it reducing the potential of drug-associated cues to elicit opiate craving and subsequent relapse. This potential protection from relapse is important in light of the current opiate crisis. However, it must be cautioned that the findings to date only suggest an effect of CBD on craving and cue-induced relapse. Long-term human randomized clinical trial studies are necessary to provide confidence that CBD may be used as a treatment for opiate dependence.

There is mixed evidence of the potential of CBD to treat use disorders other than opiate use disorder. CBD does not appear to be effective in treating relapse or cocaine craving among people being treated for cocaine use disorder, even though there is some preclinical evidence for its efficacy in this disorder. The very limited evidence of the potential of CBD to reduce nicotine intake suggested that inhaled CBD may reduce the number of cigarettes smoked, but without a reduction in craving or withdrawal. Preclinical evidence suggests that high doses of CBD may reduce alcohol intake and alcohol seeking in rodents; however, there have been no high-quality human clinical trials demonstrating efficacy of CBD for this disorder. Preclinical evidence also suggests that CBD may reduce seizures in rodents undergoing withdrawal from excessive alcohol consumption as well as alcohol-related liver damage, which can lead to hepatitis and cirrhosis and even cognitive deficits seen due to damage to the brain. However, the clinical evidence for such effects is missing, and these results must be interpreted with caution until such studies are conducted in humans. Finally, although there is limited evidence in humans that CBD may be useful in treating cannabis use disorder, most of the evidence to date is based on nabiximols treatment in which participants are treated with a combination of THC and CBD. Ongoing clinical trials suggest that nabiximols may be useful for this latter indication, but we are still awaiting the results of high-quality clinical trials.

# 12
# Conclusion

We have reviewed the evidence for the therapeutic effectiveness of CBD for several disease states. It is clear that the evidence for some indications is of much higher quality than for others. In chapter 1, we outline the various levels of evidence required by the FDA for the approval of a drug as a treatment for a disease (see figure 1.2). The treatment is first evaluated for in vitro effectiveness in reducing cellular signals indicative of the disease state. With positive results, it is then brought to preclinical trials, using animal models of the disease state. Only after sufficient preclinical evidence is obtained can a drug be brought to human clinical trials for testing, with randomized, double-blind, controlled studies as the final proof of efficacy of the drug. For some indications described in the book, the evidence is available only at the in vitro level with no preclinical in vivo evidence. We consider this the lowest level of evidence that CBD will be effective for that indication because the effect seen in the petri dish has not even been demonstrated in a whole living organism, much less tested in humans in well-controlled trials. Some indications have been shown to be effective in both experimental in vitro and in vivo preclinical animal models of the disease but not by testing in human clinical trials. We consider this to be intermediate level of evidence, because CBD has not been experimentally demonstrated to be effective against that disease in human patients. Some indications have, however, been assessed in randomized, controlled human trials and have been found to be effective. We consider this to be a high level of evidence because CBD has been experimentally shown to produce its therapeutic effect in human patients. Finally, we give the gold star of approval to the evidence supporting Epidiolex for treatment of certain types of childhood epilepsy, which has not only been shown to be effective in several experimental randomized, placebo-controlled trials in human patients but also has received FDA approval for this indication. Here we summarize the current state of evidence for the efficacy of CBD as a treatment for specific conditions.

## FDA-Approved Use of CBD: Evidence Based on High-Quality Randomized, Double-Blind, Controlled Human Trials

Clearly the highest-quality evidence for the effectiveness of CBD as a treatment is for alleviation of seizures in children with Dravet syndrome, Lennox-Gesasut syndrome and tuberous sclerosis complex. These are the only indications for which pure CBD (Epidiolex, GW Pharmaceuticals) has received FDA approval following several recent rigorous randomized, double-blind, control trials, published in the high impact *New England Journal of Medicine*. CBD has not received FDA approval as a treatment for any other indications.

## Evidence Based on Very Limited Randomized, Double-Blind, Controlled Trials

There are a few indications for which CBD has been found to be effective in randomized, double blind, placebo-controlled human trials, but has not yet been approved by the FDA for the condition. We consider this to be limited evidence as there is only somewhat promising evidence for the efficacy of CBD for these conditions.

**Opiate addiction:** Oral CBD (Epidiolex at doses of 400 mg and 800 mg) was recently evaluated in a double-blind, randomized, placebo-controlled trial to examine the effects of acute and short-term CBD administration on craving and anxiety in abstinent men and women with heroin use disorder. Both doses of CBD reduced cue-induced heroin craving, anxiety, and physiological measures of stress reactivity, with long-lasting effects up to a week later. These findings are hopeful for the treatment of craving, a classic contributor to heroin relapse in abstinent users; however, the findings do not indicate that CBD will reduce all aspects of opiate addiction in this population. There are current ongoing clinical trials listed on https://govtrials.org for CBD as a treatment for opioid, alcohol, and cannabis use disorders.

**Anxiety:** There is considerable evidence from randomized clinical trials that CBD may be an effective treatment for anxiety. Randomized, double-blind, placebo-controlled trials have shown that CBD (oral 300–600 mg) reduced anxiety in adults and teenagers with social anxiety disorder, psychiatric patients at high risk for psychosis, and Parkinson's disease patients. There are currently several ongoing clinical trials of the efficacy of CBD as a treatment for anxiety in several disease states. Anxiety is therefore a promising target disease for CBD.

Several case studies suggest that CBD may also be an effective treatment for PTSD. There are no published RCTs to date that have been conducted to evaluate the efficacy of CBD alone as a treatment for PTSD. However, a well-controlled RCT with smoked high CBD/low THC cannabis did not show improvement relative to placebo in PTSD symptoms (Bonn-Miller et al, 2021). Several ongoing and planned clinical trials for CBD alone and PTSD are listed on www.clinicaltrials.gov. Preclinical animal research suggests that CBD is much more effective in animals that are prestressed than in nonstressed animals, suggesting that it may be more effective in treating anxiety in PTSD patients than in normal controls. However, this is not the quality of evidence required to be considered of high quality. There are also no published properly controlled clinical trials on the effectiveness of CBD on patients with major depressive disorder.

**Schizophrenia:** CBD has shown some promise as a potential treatment for the positive psychotic symptoms and possibly the cognitive symptoms, but not the negative symptoms, of schizophrenia. A phase 2 randomized, double-blind, parallel group design compared CBD treatment with that of a standard antipsychotic drug, amisulpride. CBD (escalating dose from 200 mg/day to 800 mg/day over four weeks) was as effective as amilsulpride, and with a superior side effect profile. However, using a lower dose of CBD (600 mg/day over a six-week period), a double blind, placebo-controlled trial revealed that CBD did not improve positive or cognitive psychotic symptoms. The lack of effect may have been dose related, because a double-blind, placebo-controlled, clinical trial in which patients received a higher dose of CBD (1 g/day for six weeks), in addition to their typical antipsychotic medication, revealed lower levels of positive psychotic symptoms and improved illness severity and patient functioning scores, but CBD did not affect their cognitive performance. Most recently, a double-blind, placebo-controlled trial with a high dose of CBD (up to 800 mg/day), administered as a monotherapy treatment (without other antipsychotic treatments), revealed an improvement in both positive symptoms and some cognitive symptoms. The potential of CBD as a treatment for schizophrenia has generated considerable recent clinical trials. The website www.clinicaltrials.gov currently lists sixteen human clinical trials to assess the efficacy of CBD for schizophrenia. The publication of the findings of these trials will provide a better understanding of whether CBD is an effective treatment for psychotic disorders.

**Irritable bowel disorder (IBD):** Five randomized, double-blind placebo-controlled trials have evaluated the effect of CBD on intestinal permeability and inflammation in humans. In the first trial, CBD (600 mg, oral) prevented aspirin-induced inflammation and permeability changes in thirty healthy volunteers. Two of the RCTs were conducted in IBD patients and two in Crohn's disease patients, with mixed results that are likely related to dosing. The first study found that a low dose of CBD (10 mg, oral twice daily for eight weeks) was safe but ineffective in improving Crohn's disease. However, the other two trials found that a higher dose of CBD (50 mg, oral capsule, twice a day) or a CBD-rich cannabis oil reduced illness severity and improved quality-of-life outcomes, but did not affect remission rates. Two clinical trials are currently listed on www.clinicaltrials.gov with one trial evaluating CBD as a treatment for IBD and one trial evaluating CBD as a treatment for Crohn's disease.

**Graft versus host disease:** Promising human clinical trial data suggest that CBD may decrease the autoimmune reaction of graft versus host disease in transplant patients.

**Pain:** The final and least convincing indication for efficacy in human clinical trials is that of pain. Despite an abundance of clear evidence that CBD reduces pain in several preclinical animal models, the evidence that CBD reduces pain in humans is less convincing. Only a handful of participants have been tested, and the majority of these trials were with transdermal administration of CBD, but these studies did not measure the accumulated levels of the administered CBD across the skin. It is surprising that there are very few properly conducted controlled studies for the efficacy of CBD to reduce pain in humans, because considerable anecdotal evidence exists for this indication. CBD alone has been investigated in few published human clinical trials, with mixed results. A double-blind, randomized, placebo-controlled crossover trial with twenty multiple sclerosis patients unresponsive to other pain treatments was conducted with two-week treatment periods. CBD (22.5 mg/day, sublingual spray) reduced pain and spasticity over the treatment period. This was limited by a very small sample size. Two more recent studies have administered CBD by topical application for pain relief. A double-blind, placebo-controlled study determined the effect of transdermal CBD (ointment with 20 percent CBD oil, 70 mg CBD/ml, twice daily for fourteen days) in sixty patients with myofascial pain. CBD reduced pain intensity ratings with no adverse effects. Another four-week, placebo-controlled crossover trial with CBD oil (2.8 mg/ml for four weeks) in twenty-nine neuropathic pain patients reduced pain ratings even at this much lower concentration. Current ongoing clinical trials appearing on www.clinicaltrials.gov are investigating the interaction of CBD and morphine effects on pain sensitivity (NCT04030442).

**Conclusion:** To date, the only indications for which CBD is FDA approved are childhood epilepsy in Dravet syndrome, Lennox-Gastaut syndrome, and tuberous sclerosis complex. However, several other uses have been the subject of RCTs, including opiate addiction, anxiety, schizophrenia, pain, irritable bowel syndrome, and graft versus host disease. To our knowledge, there have been no randomized placebo-controlled trials of the efficacy of CBD to treat any other indication described in this book, including disorders of the skin, cancer, neuroprotective effects, neurodegenerative diseases, or nausea and vomiting.

## Evidence Based on Preclinical in Vivo Animal Models

High-equality evidence of the effectiveness of CBD in treating human health conditions comes from RCTs. However, there are some indications of CBD efficacy that extend beyond in vitro testing to preclinical in vivo testing in animal models of the disease.

Such in vivo evidence holds promise for future evaluation of CBD treatment for that disease in human clinical trials, suggesting that given the regulatory changes in the ease of conducting research with CBD, some of these indications are likely to be subjected to RCTs in the coming years.

**Pain:** Despite the rather low-quality RCT evidence of the effectiveness of CBD to reduce pain, there is considerable preclinical evidence that CBD and CBDA alone reduce pain in several animal models. CBD does not appear to have an analgesic effect in animal models of acute pain but does potentiate the analgesic effect of THC and morphine in these models. However, some synthetic analogs of CBD do exert effects alone in these acute pain models.

In animal models of neuropathic pain, including chronic or partial constriction injury of the sciatic nerve, both CBD and CBDA methyl ester reduced thermal pain sensitivity, but the latter effect was more potent. CBD alone and combined with low doses of THC prevents or reduces chemotherapy-induced pain sensitivity, suggesting it may be an effective adjunct treatment for chemotherapy-induced pain. CBD has also shown promise in reducing sensitivity to pain in diabetic mice and in a model of postoperative pain.

Inflammatory pain, induced by irritant injection into a rodent's paw, produces inflammation and physical swelling (edema) as seen in tissue injury such as a sprain. Systemically administered CBD and CBDA reduced inflammatory pain and swelling in animal models. This is a product of its anti-inflammatory activity. CBD is also a good candidate for treatment of arthritic joints based on the preclinical animal evidence. Several clinical trials are currently listed on www.clinicaltrials.gov for CBD as a treatment for pain.

**Stroke (cerebral ischemia):** CBD has been shown to be neuroprotective against stroke in several preclinical models. In an in vivo rat stroke model, CBD pretreated rats were significantly less impaired in neurological tests and showed significantly (60 percent) reduced brain damage from the stroke. Subsequent research with gerbils found that CBD could also be given five minutes after the stroke to prevent damage and that tolerance did not develop to the neuroprotective effects of CBD over fourteen days of treatment. Furthermore, when administered thirty minutes after hypoxic ischemic brain injury in newborn pigs, CBD produced a recovery of brain activity during the six-hour observation period and reduced the number of damaged neurons in the brain. When combined with hypothermia (the currently approved treatment), CBD enhanced the anti-inflammatory effects (reducing the increase in TNF-$\alpha$), reduced neuronal death, and restored brain activity in newborn pigs. The therapeutic window for CBD's beneficial effects seems to extend to eighteen hours following hypoxic ischemic brain injury. Taken together, these findings suggest that CBD administration (within eighteen hours after hypoxia ischemia) as an adjunct treatment to hypothermia is neuroprotective in animal models of hypoxic ischemic brain injury. These findings are promising for future human clinical trials of the efficacy of CBD to restore function in cases of cerebral ischemia.

**Heart and liver dysfunction:** CBD may have neuroprotective effects against heart dysfunction and in liver disease. It has been found to be neuroprotective against dysfunction of the heart in mouse models of chemotherapy-induced cardiomyopathy and type 1 diabetes–induced myocardial dysfunction. These findings provided the rationale for an exploratory human clinical trial to assess the safety of CBD in patients with heart failure (NCT03634189). In a mouse model of hepatic ischemia and alcohol binge-induced liver injury, CBD reduced liver inflammation and cell death and improved cognitive dysfunction in a mouse model of hepatic encephalopathy produced by bile duct ligation. These preclinical findings may provide the proof of principle to proceed to human clinical testing for these potential neuroprotective effects of CBD.

**Neurodegenerative disorders:** Although there is no human clinical trial evidence that CBD is effective in treating the neurodegenerative disorders of multiple sclerosis, Alzheimer's disease, Parkinson's disease, or Huntington's disease, there is promising preclinical evidence of its potential effectiveness in treating these disorders. In the case of multiple sclerosis, a recent preclinical study with mice suggests that daily treatment with a 1 percent pure CBD cream in propylene glycol may exert neuroprotective effects against autoimmune encephalitis, which is characterized by paralysis of the hind limbs, used as a model of MS. In a mouse model of Alzheimer's disease, administration of CBD for three weeks was shown to prevent amyloid-$\beta$-induced microglia activation in vivo and reduce memory impairments produced by amyloid-$\beta$. The preclinical evidence for the effectiveness of CBD to restore dopamine levels in rodent models of Parkinson disease is very limited, with one study showing that CBD restored dopamine levels but another showing it was ineffective. In a recent, single open label, dose escalation study in human Parkinson's disease patients to determine the tolerability and efficacy of oral CBD (Epidiolex), CBD was titrated from 5 to 25 mg/kg/day and maintained for fifteen days. Most patients reported some improvement in motor symptoms, but also reported mild side effects. Finally, early clinical trials with CBD aimed at treating dysfunction in Huntington's disease were without success despite in vitro evidence that CBD may be neuroprotective in this disease by acting as an antioxidant, thereby decreasing oxidative injury in HD. To the best of our knowledge, there is no preclinical evidence for the efficacy of CBD in treating Huntington's disease.

**Autoimmune diseases:** Preclinical animal studies suggest that CBD may be a useful treatment in autoimmune diseases such as diabetes type 1, hepatitis, and myocarditis. CBD lowered the incidence of diabetes in young nonobese diabetes-prone mice and reduced associated pro-inflammatory cytokine levels. CBD also reduced experimental hepatitis and its associated inflammatory mediators in mice. In addition, chronic CBD treatment attenuated the inflammatory response and cardiac dysfunction in a mouse model of autoimmune myocarditis. CBD has been shown to be effective in several preclinical models of autoimmune diseases. Hence, it should be tested in other such diseases.

**Melanoma:** In a mouse model of melanoma, CBD was more effective than placebo in reducing the tumor growth rate, but the chemotherapy drug cisplatin was more effective than CBD. Cisplatin, however, produced severe side effects, unlike CBD. This finding suggests that CBD may be tested as an adjunctive treatment against melanoma in the future.

**Other cancers:** The potential efficacy of CBD to treat other cancers is linked to its unique ability to target multiple cellular pathways controlling tumorigenesis depending on cancer type. There is considerable in vitro evidence of CBD's ability to inhibit cancer cell growth but there is limited in vivo preclinical evidence. CBD was shown to reduce glioma tumor growth in vivo in immune-deficient mice implanted with human glioma cells. CBD was also effective in vivo in reducing the primary tumor mass and the size and number of metastatic foci in a mouse model of breast cancer. Additionally, in vivo studies in thymic aplastic nude mice revealed a significant inhibition of lung metastases following CBD treatment. In a preclinical mouse model of colon cancer, CBD reduced polyps and tumors. This in vitro and in vivo evidence of CBD's antitumor properties may warrant further testing.

**Nausea and vomiting:** Although there have been no human clinical trials on the potential of CBD alone to reduce nausea and vomiting, there is considerable in vivo preclinical evidence of the therapeutic potential of CBD for these indications. Low doses (1–10 mg/kg, i.p.) of CBD reduce lithium and cisplatin-induced vomiting in shrews, but high doses (20–40 mg/kg, i.p.) of CBD increase such vomiting. CBD also reduces lithium-induced acute nausea in rats, as assessed by the selective gaping model, but it is effective across a dose range of 1–20 mg/kg, i.p. CBDA is about 1,000 times more potent in reducing nausea and vomiting than CBD and synergistically acts with the classic antinausea compound ondansetron to reduce nausea in rats. The more stable CBDA methyl ester (HU-580) is even more potent against nausea than CBDA. As well, unlike the typically prescribed antinausea drug ondansetron, CBD and CBDA interfere with anticipatory nausea (nausea in anticipation of chemotherapy treatments) in a rat model. These results beg for a clinical trial with HU-580, CBDA, and CBD to attenuate nausea and vomiting in chemotherapy.

**Posttraumatic stress disorder:** Preclinical evidence suggests that CBD reduces the formation, expression, and reconsolidation of fear memories. Importantly, it also facilitates the extinction of these fear memories, making it a promising therapeutic agent for PTSD patients. In animal models, CBD appears much more effective in reducing anxiety in animals that are previously stressed than in those not previously stressed, which may be particularly important for PTSD patient treatments. In line with this evidence, a few case reports and retrospective studies suggest that CBD may be beneficial in alleviating some symptoms of PTSD such as nightmares. Four clinical trials are underway examining the effects of CBD treatment on PTSD patients.

**Depression:** CBD exerts antidepressant effects in a number of preclinical models, and initial case studies also show that it may be beneficial as a treatment for depression. Properly controlled clinical trials are needed to investigate these potential antidepressant effects and determine if CBD's effects can be maintained with chronic treatment. No such trials are currently registered.

**Addiction:** Addiction is a chronic disorder, such that drug-associated environmental cues trigger relapse to drug use. Human clinical trial data show great promise for CBD to prevent heroin cue-induced relapse in human opiate addiction. There is also human trial evidence that CBD may reduce nicotine seeking but not necessarily craving. In the case of alcohol addiction, preclinical evidence indicates that CBD may reduce alcohol consumption and cue- and stress-induced relapse, and potentially protect against certain harmful effects of alcohol, including liver and brain damage. CBD may also reduce alcohol withdrawal reactions in mice. Finally, at high doses and repeated treatments, CBD may reduce stimulant self-administration in animal models.

**Conclusion:** Preclinical evidence in animal models is a necessary first step that ultimately leads to high-quality RCTs for the search of a treatment for a disease condition. Considerable preclinical evidence exists for CBD as a treatment for inflammatory and neuropathic, but not acute, pain. CBD also has promise in neuroprotection against stroke, heart, and liver dysfunction in animal models and may be a useful treatment in autoimmune diseases. There is also limited preclinical evidence that CBD may reduce symptoms of multiple sclerosis and Alzheimer's disease. There is also limited preclinical evidence that CBD may reduce tumor growth in several cancers, including melanoma, breast cancer, and gliomas. CBD and CBDA clearly act as antiemetic (within a limited dose range) drugs in shrews and as antinausea drugs in rats. Preclinical models of PTSD, depression, and addiction reveal promise for CBD as a treatment for these conditions. These findings show promise of treatments with CBD, but one cannot conclude that humans will respond in the same manner until RCTs are conducted for these indications.

## Evidence Based Only on In Vitro Tests

Most removed from human RCT evidence is that derived only from in vitro tests in the laboratory. CBD has been shown to modify signals in cells for several disease states for which no other evidence is available, including animal testing. This is the first level of evidence often leading to subsequent preclininal and clinical testing. Often the lack of animal testing is a function of the lack of a relevant animal model for that condition, such as acne vulgaris and psoriasis. Most of the touted cosmetic effects of CBD for skin products fall in this very low quality of evidence category. However, four human trials are listed on www.clinicaltrials.gov for treatment of skin with CBD for hydration, acne, and dermatitis. Evidence that CBD may be an effective treatment for psoriasis and acne is solely based on such in vitro evidence, and generalization of such evidence to the whole organism must be taken cautiously until RCTs are conducted.

## Where Does the Science Go from Here?

There is obviously an abundance of in vitro and preclinical in vivo evidence for the efficacy of CBD to treat many human medical conditions but few human clinical trials to support this evidence. It is critical that these claims are evaluated in high-quality human RCTs. As it stands, many of the purported effects of CBD remain in the realm of hype until such trials are completed to support or refute many of the claimed benefits of CBD. Are the claimed benefits merely placebo effects based on expectation, as has been reported recently with antianxiety effects (Spinella et al. 2021), or are they real clinical effects? Only RCTs will provide that answer.

This book provides an overview of the indications for which there is some basic scientific grounding on the potential medicinal benefit claims of CBD. This may represent at least a starting point at separating the hype from the hope of this very interesting cannabinoid compound that has relatively little toxicity (with the caution of interactions with drug metabolizing liver enzymes) and does not produce the THC "high."

# References

Abame, M. A., Y. He, S. Wu, Z. Xie, J. Zhang, X. Gong, C. Wu, and J. Shen. 2021. "Chronic administration of synthetic cannabidiol induces antidepressant effects involving modulation of serotonin and noradrenaline levels in the hippocampus." *Neurosci Lett* 744: 135594. doi:10.1016/j.neulet.2020.135594.

Abraham, A. D., E. J. Y. Leung, B. A. Wong, Z. M. G. Rivera, L. C. Kruse, J. J. Clark, and B. B. Land. 2019. "Orally consumed cannabinoids provide longlasting relief of allodynia in a mouse model of chronic neuropathic pain." *Neuropsychopharmacology* 45: 1105–1114. doi:10.1038/s41386-019-0585-3.

Adams, R., M. Hunt, and J. H. Clark. 1940. "Structure of cannabidiol, a product isolated from the marihuana extract of Minnesota wild hemp." *J Amer Chem Soc* 62: 196–200.

Agurell, S., S. Carlsson, J. E. Lindgren, A. Ohlsson, H. Gillespie, and L. Hollister. 1981. "Interactions of delta 1-tetrahydrocannabinol with cannabinol and cannabidiol following oral administration in man: Assay of cannabinol and cannabidiol by mass fragmentography." *Experientia* 37 (10): 1090–1092. doi:10.1007/bf02085029.

Ahmed, A., M. A. van der Marck, G. van den Elsen, and M. Olde Rikkert. 2015. "Cannabinoids in late-onset Alzheimer's disease." *Clinical Pharmacology and Therapeutics* 97: 597–606.

Ahmed, S. A., S. A. Ross, D. Slade, M. M. Radwan, F. Zulfiqar, R. R. Matsumoto, Y. T. Xu, E. Viard, R. C. Speth, V. T. Karamyan, and M. A. ElSohly.

2008. "Cannabinoid ester constituents from high-potency *Cannabis sativa*." *J Nat Prod* 71 (4): 536–542. doi:10.1021/np070454a.

Alhamoruni, A., A. C. Lee, K. L. Wright, M. Larvin, and S. E. O'Sullivan. 2010. "Pharmacological effects of cannabinoids on the Caco-2 cell culture model of intestinal permeability." *J Pharmacol Exp Ther* 335 (1): 92–102. doi:10.1124/jpet.110.168237.

Alhamoruni, A., K. L. Wright, M. Larvin, and S. E. O'Sullivan. 2012. "Cannabinoids mediate opposing effects on inflammation-induced intestinal permeability." *Br J Pharmacol* 165 (8): 2598–610. doi:10.1111/j.1476-5381.2011.01589.x.

Allen, J. H., G. M. de Moore, R. Heddle, and J. C. Twartz. 2004. "Cannabinoid hyperemesis: Cyclical hyperemesis in association with chronic cannabis abuse." *Gut* 53 (11): 1566–1570. doi:10.1136/gut.2003.036350.

Allsop, D. J., J. Copeland, N. Lintzeris, A. J. Dunlop, M. Montebello, C. Sadler, G. R. Rivas, R. M. Holland, P. Muhleisen, M. M. Norberg, J. Booth, and I. S. McGregor. 2014. "Nabiximols as an agonist replacement therapy during cannabis withdrawal: A randomized clinical trial." *JAMA Psychiatry* 71 (3): 281–291. doi:10.1001/jamapsychiatry.2013.3947.

Almeida, V., R. Levin, F. F. Peres, S. T. Niigaki, M. B. Calzavara, A. W. Zuardi, J. E. Hallak, J. A. Crippa, and V. C. Abilio. 2013. "Cannabidiol exhibits anxiolytic but not antipsychotic property evaluated in the social interaction test." *Prog Neuropsychopharmacol Biol Psychiatry* 41: 30–35. doi:10.1016/j.pnpbp.2012.10.024.

Alvarez, F. J., H. Lafuente, M. C. Rey-Santano, V. E. Mielgo, E. Gastiasoro, M. Rueda, R. G. Pertwee, A. I. Castillo, J. Romero, and J. Martinez-Orgado. 2008. "Neuroprotective effects of the nonpsychoactive cannabinoid cannabidiol in hypoxic-ischemic newborn piglets." *Pediatr Res* 64 (6): 653–658. doi:10.1203/PDR.0b013e318186e5dd.

Ames, F. R., and S. Cridland. 1986. "Anticonvulsant effect of cannabidiol." *S Afr Med J* 69 (1): 14.

Anderson, L. L., N. L. Absalom, S. V. Abelev, I. K. Low, P. T. Doohan, L. J. Martin, M. Chebib, I. S. McGregor, and J. C. Arnold. 2019. "Coadministered cannabidiol and clobazam: Preclinical evidence for both pharmacodynamic and pharmacokinetic interactions." *Epilepsia* 60 (11): 2224–2234. doi:10.1111/epi.16355.

Andries, A., J. Frystyk, A. Flyvbjerg, and R. K. Støving. 2014. "Dronabinol in severe, enduring anorexia nervosa: A randomized controlled trial." *Int J Eat Disord* 47 (1): 18–23. doi:10.1002/eat.22173.

Andries, A., J. Frystyk, A. Flyvbjerg, and R. K. Støving. 2015. "Changes in IGF-I, urinary free cortisol and adipokines during dronabinol therapy in anorexia nervosa: Results from a randomised, controlled trial." *Growth Horm IGF Res* 25 (5): 247–252. doi:10.1016/j.ghir.2015.07.006.

Anil, S. M., N. Shalev, A. C. Vinayaka, S. Nadarajan, D. Namdar, E. Belausov, I. Shoval, K. A. Mani, G. Mechrez, and H. Koltai. 2021. "Cannabis compounds exhibit anti-inflammatory activity in vitro in COVID-19-related inflammation in lung epithelial cells and pro-inflammatory activity in macrophages." *Sci Rep* 11 (1): 1462. doi:10.1038/s41598-021-81049-2.

Appiah-Kusi, E., N. Petros, R. Wilson, M. Colizzi, M. G. Bossong, L. Valmaggia, V. Mondelli, P. McGuire, and S. Bhattacharyya. 2020. "Effects of short-term cannabidiol treatment on response to social stress in subjects at clinical high risk of developing psychosis." *Psychopharmacology (Berl)* 237: 1121–1130. doi:10.1007/s00213-019-05442-6.

Arciniegas, D. B. 2015. "Psychosis." *Continuum (Minneap Minn)* 21 (3 Behavioral Neurology and Neuropsychiatry): 715–736. doi:10.1212/01.CON.0000466662.89908.e7.

Arkell, T. R., R. C. Kevin, J. Stuart, N. Lintzeris, P. S. Haber, J. G. Ramaekers, and I. S. McGregor. 2019. "Detection of delta(9) THC in oral fluid following vaporized cannabis with varied cannabidiol (CBD) content: An evaluation of two point-of-collection testing devices." *Drug Test Anal* 11 (10): 1486–1497. doi:10.1002/dta.2687.

Arkell, T. R., N. Lintzeris, R. C. Kevin, J. G. Ramaekers, R. Vandrey, C. Irwin, P. S. Haber, and I. S. McGregor. 2019. "Cannabidiol (CBD) content in vaporized cannabis does not prevent tetrahydrocannabinol (THC)-induced impairment of driving and cognition." *Psychopharmacology (Berl)* 236 (9): 2713–2724. doi:10.1007/s00213-019-05246-8.

Arndt, D. L., and H. de Wit. 2017. "Cannabidiol does not dampen responses to emotional stimuli in healthy adults." *Cannabis Cannabinoid Res* 2 (1): 105–113. doi:10.1089/can.2017.0014.

Arout, C. A., M. Haney, E. S. Herrmann, G. Bedi, and Z. D. Cooper. 2021. "The dose-dependent analgesic effects, abuse liability, safety and tolerability of oral cannabidiol in healthy humans." *Br J Clin Pharmacol* doi:10.1111 /bcp.14973.

Ashton, C. H., P. B. Moore, P. Gallagher, and A. H. Young. 2005. "Cannabinoids in bipolar affective disorder: A review and discussion of their therapeutic potential." *J Psychopharmacol* 19 (3): 293–300. doi:10.1177 /0269881105051541.

Aso, E., A. Sanchez-Pla, E. Vegas-Lozano, R. Maldonado, and I. Ferrer. 2015. "Cannabis-based medicine reduces multiple pathological processes in AbetaPP/PS1 mice." *J Alzheimer's Dis* 43 (3): 977–991. doi:10.3233 /JAD-141014.

Atalay, S., A. Gegotek, A. Wronski, P. Domigues, and E. Skrzydlewska. 2021. "Therapeutic application of cannabidiol on UVA and UVB irradiated rat skin. A proteomic study." *J Pharm Biomed Anal* 192: 113656. doi:10.1016/j .jpba.2020.113656.

Atsmon, J., D. Heffetz, L. Deutsch, F. Deutsch, and H. Sacks. 2018. "Single-dose pharmacokinetics of oral cannabidiol following administration of PTL101: A new formulation based on gelatin matrix pellets technology." *Clin Pharmacol Drug Dev* 7 (7): 751–758. doi:10.1002/cpdd.408.

Aviram, J., G. M. Lewitus, Y. Vysotski, A. Uribayev, S. Procaccia, I. Cohen, A. Leibovici, M. Abo-Amna, L. Akria, D. Goncharov, N. Mativ, A. Kauffman, A. Shai, O. Hazan, G. Bar-Sela, and D. Meiri. 2020. "Short-term medical cannabis treatment regimens produced beneficial effects among palliative cancer patients." *Pharmaceuticals* 13 (12): 435. doi:10.3390/ph13120435.

Aymerich, M. S., E. Aso, M. A. Abellanas, R. M. Tonlon, Ramos J. A., I. Ferrer, J. Romero, and J. Fernandez-Ruiz. 2018. "Cannabinoid pharmacology/therapeutics in chronic degenerative disorders affecting the nervous system." *Biochemical Pharmacology* 157: 67–84. doi:10.1016/j.bcp.2018.08.016.

Baek, S. H., M. Srebnik, and R. Mechoulam. 1985. "Borobtrifluoride on alumina-a modified Lewis acid reagent. An improved synthesis of cannabidiol." *Tetrahedron Lett* 26:1083–1086.

Balachandran, P., M. Elsohly, and K. P. Hill. 2021. "Cannabidiol interactions with medications, illicit substances, and alcohol: A comprehensive review." *J Gen Intern Med* 36: 2074–2084. doi:10.1007/s11606-020-06504-8.

Bandelow, B., S. Michaelis, and D. Wedekind. 2017. "Treatment of anxiety disorders." *Dialogues in Clinical Neuroscience* 19 (2): 93–107. doi:10.31887 /DCNS.2017.19.2/bbandelow.

Bandelow, B., U. Seidler-Brandler, A. Becker, D. Wedekind, and E. Rüther. 2007. "Meta-analysis of randomized controlled comparisons of psychopharmacological and psychological treatments for anxiety disorders." *World Journal of Biological Psychiatry.* 8 (3): 175–187. doi:10.1080/15622970601110273.

Bar-Sela, G., D. Zalman, V. Semenysty, and E. Ballan. 2019. "The effects of dosage-controlled cannabis capsules on cancer-related cachexia and anorexia syndrome in advanced cancer patients: Pilot study." *Integr Cancer Ther* 18: 1534735419881498. doi:10.1177/1534735419881498.

Barata, L., L. Arruza, M. J. Rodriguez, E. Aleo, E. Vierge, E. Criado, E. Sobrino, C. Vargas, M. Ceprian, A. Gutierrez-Rodriguez, W. Hind, and J. Martinez-Orgado. 2019. "Neuroprotection by cannabidiol and hypothermia in a piglet model of newborn hypoxic-ischemic brain damage." *Neuropharmacology* 146: 1–11. doi:10.1016/j.neuropharm.2018.11.020.

Baron, E. P., P. Lucas, J. Eades, and O. Hogue. 2018. "Patterns of medicinal cannabis use, strain analysis, and substitution effect among patients with migraine, headache, arthritis, and chronic pain in a medicinal cannabis cohort." *J Headache Pain* 19 (1): 37. doi:10.1186/s10194-018-0862-2.

Bartner, L. R., S. McGrath, S. Rao, L. K. Hyatt, and L. A. Wittenburg. 2018. "Pharmacokinetics of cannabidiol administered by 3 delivery methods at 2 different dosages to healthy dogs." *Can J Vet Res* 82 (3): 178–183.

Beale, C., S. J. Broyd, Y. Chye, C. Suo, M. Schira, P. Galettis, J. H. Martin, M. Yucel, and N. Solowij. 2018. "Prolonged cannabidiol treatment effects on hippocampal subfield volumes in current cannabis users." *Cannabis Cannabinoid Res* 3 (1): 94–107. doi:10.1089/can.2017.0047.

Bebee, B., D. M. Taylor, E. Bourke, K. Pollack, L. Foster, M. Ching, and A. Wong. 2021. "The CANBACK trial: A randomised, controlled clinical trial of oral cannabidiol for people presenting to the emergency department with acute low back pain." *Med J Aust* 214 (8): 370–375. doi:10.5694/mja2.51014.

Belendiuk, K. A., K. A. Babson, R. Vandrey, and M. O. Bonn-Miller. 2015. "Cannabis species and cannabinoid concentration preference among sleep-disturbed medicinal cannabis users." *Addict Behav* 50: 178–181. doi:10.1016/j.addbeh.2015.06.032.

Belgrave, B. E., K. D. Bird, G. B. Chesher, D. M. Jackson, K. E. Lubbe, G. A. Starmer, and R. K. Teo. 1979. "The effect of cannabidiol, alone and in combination with ethanol, on human performance." *Psychopharmacology (Berl)* 64 (2): 243–246. doi:10.1007/bf00496070.

Bergmann, K. T., K. Broekhuizen, and G. J. Groeneuveld. 2020. "Clinical trial simulations of the interaction between cannabidiol and clobazam and effect on drop-seizure frequency." *Br J Clin Pharmacol* 86 (2): 380–385. doi: 10.1111/bcp.14158.

Bergamaschi, M. M., R. H. Queiroz, M. H. Chagas, D. C. de Oliveira, B. S. De Martinis, F. Kapczinski, J. Quevedo, R. Roesler, N. Schroder, A. E. Nardi, R. Martin-Santos, J. E. Hallak, A. W. Zuardi, and J. A. Crippa. 2011. "Cannabidiol reduces the anxiety induced by simulated public speaking in treatment-naive social phobia patients." *Neuropsychopharmacology* 36 (6): 1219–1226. doi:10.1038/npp.2011.6.

Bergamaschi, M. M., R. H. Queiroz, A. W. Zuardi, and J. A. Crippa. 2011. "Safety and side effects of cannabidiol, a cannabis sativa constituent." *Curr Drug Saf* 6 (4): 237–249. doi:10.2174/157488611798280924.

Bhargava, H. N. 1976. "Effect of some cannabinoids on naloxone-precipitated abstinence in morphine dependent mice." *Psychopharmacology* 49: 267–270. doi:10.1007/BF00426828.

Bhaskar, A., A. Bell, M. Boivin, W. Briques, M. Brown, H. Clarke, C. Cyr, E. Eisenberg, R. F. de Oliveira Silva, e. Frohlich, P. Georgius, M. Hogg, T. I. Horsted, C. A. MacCallum, K. R. Müller-Vahl, C. O'Connell, R. Sealey, M. Seibolt, A. Sihota, B. K. Smith, D. Sulak, A. Vigano, and D. E. Moulin. 2021. "Consensus recommendations on dosing and administration of medical cannabis to treat chronic pain: Results of a modified Delphi process." *Journal of Cannabis Research* 3 (1): 22. doi:10.1186/s42238-021-00073-1

Bhattacharyya, S., J. A. Crippa, P. Allen, R. Martin-Santos, S. Borgwardt, P. Fusar-Poli, K. Rubia, J. Kambeitz, C. O'Carroll, M. L. Seal, V. Giampietro, M. Brammer, A. W. Zuardi, Z. Atakan, and P. K. McGuire. 2012. "Induction of psychosis by delta-9-tetrahydrocannabinol reflects modulation of prefrontal and striatal function during attentional salience processing." *Arch Gen Psychiatry* 69 (1): 27–36. doi:10.1001/archgenpsychiatry.2011.161.

Bhattacharyya, S., P. D. Morrison, P. Fusar-Poli, R. Martin-Santos, S. Borgwardt, T. Winton-Brown, C. Nosarti, C. M. O'Carroll, M. Seal, P. Allen, M. A. Mehta, J. M. Stone, N. Tunstall, V. Giampietro, S. Kapur, R. M. Murray, A. W. Zuardi, J. A. Crippa, Z. Atakan, and P. K. McGuire. 2010. "Opposite effects of delta-9-tetrahydrocannabinol and cannabidiol on human brain function and psychopathology." *Neuropsychopharmacology* 35 (3): 764–774. doi:10.1038/npp.2009.184.

Bhattacharyya, S., R. Wilson, E. Appiah-Kusi, A. O'Neill, M. Brammer, J. Perez, R. Murray, P. Allen, M. G. Bossong, and P. McGuire. 2018. "Effect of cannabidiol on medial temporal, midbrain, and striatal dysfunction in people at clinical high risk of psychosis: A randomized clinical trial." *JAMA Psychiatry* 75 (11): 1107–1117. doi:10.1001/jamapsychiatry.2018.2309.

Billig, I., B. J. Yates, and L. Rinaman. 2001. "Plasma hormone levels and central c-Fos expression in ferrets after systemic administration of cholecystokinin." *Am J Physiol Regul Integr Comp Physiol* 281 (4): R1243–1255. doi:10.1152/ajpregu.2001.281.4.R1243.

Bird, K. D., T. Boleyn, G. B. Chesher, D. M. Jackson, G. A. Starmer, and Teo R. K. 1980. "Intercannabinoid and cannabnoid-ethanol interactions on human performance." *Psychopharmacology* 71: 181–188. doi:10.1007/BF00434409.

Birnbaum, A. K., A. Karanam, S. E. Marino, C. M. Barkley, R. P. Remmel, M. Roslawski, M. Gramling-Aden, and I. E. Leppik. 2019. "Food effect on pharmacokinetics of cannabidiol oral capsules in adult patients with refractory epilepsy." *Epilepsia* 60 (8): 1586–1592. doi:10.1111/epi.16093.

Bisogno, T., L. Hanus, L. De Petrocellis, S. Tchilibon, D. E. Ponde, I. Brandi, A. S. Moriello, J. B. Davis, R. Mechoulam, and V. Di Marzo. 2001. "Molecular targets for cannabidiol and its synthetic analogues: Effect on vanilloid VR1 receptors and on the cellular uptake and enzymatic hydrolysis of anandamide." *Br J Pharmacol* 134 (4): 845–852. doi:10.1038/sj.bjp.0704327.

Bitencourt, R. M., F. A. Pamplona, and R. N. Takahashi. 2008. "Facilitation of contextual fear memory extinction and anti-anxiogenic effects of AM404 and cannabidiol in conditioned rats." *Eur Neuropsychopharmacol* 18 (12): 849–859. doi:10.1016/j.euroneuro.2008.07.001.

Blake, D. R., P. Robson, M. Ho, R. W. Jubb, and C. S. McCabe. 2006. "Preliminary assessment of the efficacy, tolerability and safety of a cannabis-based medicine (Sativex) in the treatment of pain caused by rheumatoid arthritis." *Rheumatology (Oxford)* 45 (1): 50–52. doi:10.1093/rheumatology/kei183.

Blessing, E. M., M. M. Steenkamp, J. Manzanares, and C. R. Marmar. 2015. "Cannabidiol as a potential treatment for anxiety disorders." *Neurotherapeutics* 12 (4): 825–836. doi:10.1007/s13311-015-0387-1.

Blier, P., G. Pineyro, M. el Mansari, R. Bergeron, and C. de Montigny. 1998. "Role of somatodendritic 5-HT autoreceptors in modulating 5-HT neurotransmission." *Annals of the New York Academy of Sciences* 861: 204–216. doi:10.1111/j.1749-6632.1998.tb10192.x.

Blount, B. C., M. P. Karwowski, P. G. Shields, M. Morel-Espinosa, L. Valentin-Blasini, M. Gardner, M. Braselton, C. R. Brosius, K. T. Caron, D. Chambers, J. Corstvet, E. Cowan, V. R. De Jesus, P. Espinosa, C. Fernandez, C. Holder, Z. Kuklenyik, J. D. Kusovschi, C. Newman, G. B. Reis, J. Rees, C. Reese, L. Silva, T. Seyler, M. A. Song, C. Sosnoff, C. R. Spitzer, D. Tevis, L. Wang, C. Watson, M. D. Wewers, B. Xia, D. T. Heitkemper, I. Ghinai, J. Layden, P. Briss, B. A. King, L. J. Delaney, C. M. Jones, G. T. Baldwin, A. Patel, D. Meaney-Delman, D. Rose, V. Krishnasamy, J. R. Barr, J. Thomas, J. L. Pirkle, and Group Lung Injury Response Laboratory Working. 2020. "Vitamin E acetate in bronchoalveolar-lavage fluid associated with EVALI." *N Engl J Med* 382 (8): 697–705. doi:10.1056/NEJMoa1916433.

Bluett, R. J., J. C. Gamble-George, D. J. Hermanson, N. D. Hartley, L. J. Marnett, and S. Patel. 2014. "Central anandamide deficiency predicts stress-induced anxiety: behavioral reversal through endocannabinoid augmentation." *Transl Psychiatry* 4 (7): e408. doi:10.1038/tp.2014.53.

Boehnke, K. F., J. J. Gagnier, L. Matallana, and D. A. Williams. 2021. "Substituting cannabidiol for opioids and pain medications among individuals with fibromyalgia: A large online survey." *J Pain* 22 (11): 1418–1428. doi:10.1016/j.jpain.2021.04.011.

Boggs, D. L., J. D. Nguyen, D. Morgenson, M. A. Taffe, and M. Ranganathan. 2018. "Clinical and preclinical evidence for functional interactions of cannabidiol and delta-(9)-tetrahydrocannabinol." *Neuropsychopharmacology* 43 (1): 142–154. doi:10.1038/npp.2017.209.

Boggs, D. L., T. Surti, A. Gupta, S. Gupta, M. Niciu, B. Pittman, A. M. Schnakenberg Martin, H. Thurnauer, A. Davies, D. C. D'Souza, and M. Ranganathan. 2018. "The effects of cannabidiol (CBD) on cognition and symptoms in outpatients with chronic schizophrenia: A randomized placebo controlled trial." *Psychopharmacology (Berl)* 235 (7): 1923–1932. doi:10.1007/s00213-018-4885-9.

Bolognini, D., E. M. Rock, N. L. Cluny, M. G. Cascio, C. L. Limebeer, M. Duncan, C. G. Stott, F. A. Javid, L. A. Parker, and R. G. Pertwee. 2013. "Cannabidiolic acid prevents vomiting in *Suncus murinus* and nausea-induced behaviour in rats by enhancing 5-HT1A receptor activation." *Br J Pharmacol* 168 (6): 1456–1470. doi:10.1111/bph.12043.

Bolsoni, L. M., T. D. A. da Silva, S. M. Quintana, M. de Castro, J. A. Crippa, and A. W. Zuardi. 2019. "Changes in cortisol awakening response before and after development of posttraumatic stress disorder, which cannot be avoided with use of cannabidiol: A case report." *Perm J* 23:18.300. doi:10.7812/TPP/18.300.

Bonn-Miller, M. O., S. L. Banks, and T. Sebree. 2017. "Conversion of cannabidiol following oral administration: Authors' response to Grotenhermen et al." *Cannabis Cannabinoid Res* 2 (1): 5–7. doi:10.1089/can.2016.0038.

Bonn-Miller, M. O., M. J. E. Loflin, B. F. Thomas, J. P. Marcu, T. Hyke, and R. Vandrey. 2017. "Labeling accuracy of cannabidiol extracts sold online." *JAMA* 318 (17): 1708–1709. doi:10.1001/jama.2017.11909.

Bonn-Miller, M. O., S. Sisley, P. Riggs, B. Yazar-Klosinski, J. B. Wang, M. J. E. Loflin, B. Shechet, C. Hennigan, R. Matthews, A. Emerson, and R. Doblin. 2021. "The short-term impact of 3 smoked cannabis preparations versus placebo on PTSD symptoms: A randomized cross-over clinical trial." *PLoS One* 16 (3): e0246990. doi:10.1371/journal.pone.0246990.

Booker, L., P. S. Naidu, R. K. Razdan, A. Mahadevan, and A. H. Lichtman. 2009. "Evaluation of prevalent phytocannabinoids in the acetic acid model of visceral nociception." *Drug Alcohol Depend* 105 (1–2): 42–47. doi:10.1016/j.drugalcdep.2009.06.009.

Bornheim, L. M., and M. A. Correia. 1989. "Effect of cannabidiol on cytochrome P-450 isozymes." *Biochem Pharmacol* 38 (17): 2789–2794. doi:10.1016/0006-2952(89)90432-2.

Borrelli, F., G. Aviello, B. Romano, P. Orlando, R. Capasso, F. Maiello, F. Guadagno, S. Petrosino, F. Capasso, V. Di Marzo, and A. A. Izzo. 2009. "Cannabidiol, a safe and non-psychotropic ingredient of the marijuana plant *Cannabis sativa*, is protective in a murine model of colitis." *J Mol Med (Berl)* 87 (11): 1111–1121. doi:10.1007/s00109-009-0512-x.

Bourin, M., and M. Hascoet. 2003. "The mouse light/dark box test." *Eur J Pharmacol* 463 (1–3): 55–65. doi:10.1016/s0014-2999(03)01274-3.

Bowe, A., and R. Rosenheck. 2015. "PTSD and substance use disorder among veterans: Characteristics, service utilization and pharmacotherapy." *Journal of Dual Diagnosis* 11 (1): 22–32. doi:10.1080/15504263.2014.989653.

Bowers, M. E., and K. J. Ressler. 2015. "An overview of translationally informed treatments for posttraumatic stress disorder: Animal models of Pavlovian fear conditioning to human clinical trials." *Biol Psychiatry* 78 (5): E15–27. doi:10.1016/j.biopsych.2015.06.008.

Braida, D., S. Pegorini, M. V. Arcidiacono, G. G. Consalez, L. Croci, and M. Sala. 2003. "Post-ischemic treatment with cannabidiol prevents electroencephalographic flattening, hyperlocomotion and neuronal injury in gerbils." *Neurosci Lett* 346 (1–2): 61–64. doi:10.1016/s0304-3940(03)00569-x.

Breuer, A., C. G. Haj, M. V. Fogaca, F. V. Gomes, N. R. Silva, J. F. Pedrazzi, E. A. Del Bel, J. C. Hallak, J. A. Crippa, A. W. Zuardi, R. Mechoulam, and F. S. Guimaraes. 2016. "Fluorinated cannabidiol derivatives: Enhancement of activity in mice models predictive of anxiolytic, antidepressant and antipsychotic effects." *PLoS One* 11 (7): e0158779. doi:10.1371/journal.pone.0158779.

Brierley, D. I., J. Samuels, M. Duncan, B. J. Whalley, and C. M. Williams. 2016. "Neuromotor tolerability and behavioural characterisation of cannabidiolic acid, a phytocannabinoid with therapeutic potential for anticipatory nausea." *Psychopharmacology (Berl)* 233 (2): 243–254. doi:10.1007/s00213-015-4100-1.

Britch, S. C., S. Babalonis, and S. L. Walsh. 2021. "Cannabidiol: Pharmacology and therapeutic targets." *Psychopharmacology (Berl)* 238 (1): 9–28. doi:10.1007/s00213-020-05712-8.

Britch, S., A. Goodman, J. Wiley, A. Pondelick, and R. Craft. 2020. "Antinociceptive and immune effects of delta-9-tetrahydrocannabinol or cannabidiol in male versus female rats with persistent inflammatory pain." *J Pharmacol Exp Ther* 373 (3): 416–428. doi:10.1124/jpet.119.263319.

Britch, S. C., J. L. Wiley, Z. Yu, B. H. Clowers, and R. M. Craft. 2017. "Cannabidiol-delta(9)-tetrahydrocannabinol interactions on acute pain and locomotor activity." *Drug Alcohol Depend* 175: 187–197. doi:10.1016/j.drugalcdep.2017.01.046.

Brown, J. D. 2020. "Cannabidiol as prophylaxis for SARS-CoV-2 and COVID-19? Unfounded claims versus potential risks of medications during the pandemic." *Research in Social and Administrative Pharmacy* 17 (1): 2053. doi:10.1016/j.sapharm.2020.03.020.

Brunt, T. M., M. van Genugten, K. Honer-Snoeken, M. J. van de Velde, and R. J. Niesink. 2014. "Therapeutic satisfaction and subjective effects of different strains of pharmaceutical-grade cannabis." *J Clin Psychopharmacol* 34 (3): 344–349. doi:10.1097/JCP.0000000000000129.

Calpe-López, C., M. P. Garcia-Pardo, and M. A. Aguilar. 2019. "Cannabidiol treatment might promote resilience to cocaine and methamphetamine use disorders: A review of possible mechanisms." *Molecules* 24 (14): 2583. doi:10.3390/molecules24142583.

Calpe-López, C., A. Gasparyan, F. Navarrete, J. Manzanares, J. Miñarro, and M. A. Aguilar. 2021. "Cannabidiol prevents priming- and stress-induced reinstatement of the conditioned place preference induced by cocaine in mice." *J Psychopharmacol* 35 (7): 864–874. doi:10.1177/0269881120965952.

Campos, A. C., and F. S. Guimaraes. 2008. "Involvement of 5HT1A receptors in the anxiolytic-like effects of cannabidiol injected into the dorsolateral periaqueductal gray of rats." *Psychopharmacology (Berl)* 199 (2): 223–230. doi:10.1007/s00213-008-1168-x.

Campos, A. C., and F. S. Guimaraes. 2009. "Evidence for a potential role for TRPV1 receptors in the dorsolateral periaqueductal gray in the attenuation of the anxiolytic effects of cannabinoids." *Prog Neuropsychopharmacol Biol Psychiatry* 33 (8): 1517–1521. doi:10.1016/j.pnpbp.2009.08.017.

Capano, A., R. Weaver, and E. Burkman. 2020. "Evaluation of the effects of CBD hemp extract on opioid use and quality of life indicators in chronic pain patients: A prospective cohort study." *Postgrad Med* 132 (1): 56–61. doi:10.1080/00325481.2019.1685298.

Careaga, M. B. L., C. E. N. Girardi, and D. Suchecki. 2016. "Understanding posttraumatic stress disorder through fear conditioning, extinction and reconsolidation." *Neurosci Biobehav Rev* 71: 48–57. doi:10.1016/j.neubiorev.2016.08.023.

Carlini, E. A., and J. M. Cunha. 1981. "Hypnotic and antiepileptic effects of cannabidiol." *J Clin Pharmacol* 21 (S1): 417S–427S. doi:10.1002/j.1552-4604.1981.tb02622.x.

Casares, L., V. Garcia, M. Garrido-Rodriguez, E. Millan, J. A. Collado, A. Garcia-Martin, J. Penarando, M. A. Calzado, L. de la Vega, and E. Munoz. 2020. "Cannabidiol induces antioxidant pathways in keratinocytes by targeting BACH1." *Redox Biol* 28: 101321. doi:10.1016/j.redox.2019.101321.

Casey, S. L., N. Atwal, and C. W. Vaughan. 2017. "Cannabis constituent synergy in a mouse neuropathic pain model." *Pain* 158 (12): 2452–2460. doi:10.1097/j.pain.0000000000001051.

Ceprian, M., C. Vargas, L. Garcia-Toscano, F. Penna, L. Jimenez-Sanchez, S. Achicallende, I. Elezgarai, P. Grandes, W. Hind, M. R. Pazos, and

J. Martinez-Orgado. 2019. "Cannabidiol administration prevents hypoxia-ischemia-induced hypomyelination in newborn rats." *Front Pharmacol* 10: 1131. doi:10.3389/fphar.2019.01131.

Chagas, M. H., J. A. Crippa, A. W. Zuardi, J. E. Hallak, J. P. Machado-de-Sousa, C. Hirotsu, L. Maia, S. Tufik, and M. L. Andersen. 2013. "Effects of acute systemic administration of cannabidiol on sleep-wake cycle in rats." *J Psychopharmacol* 27 (3): 312–316. doi:10.1177/0269881112474524.

Chagas, M. H., A. L. Eckeli, A. W. Zuardi, M. A. Pena-Pereira, M. A. Sobreira-Neto, E. T. Sobreira, M. R. Camilo, M. M. Bergamaschi, C. H. Schenck, J. E. Hallak, V. Tumas, and J. A. Crippa. 2014. "Cannabidiol can improve complex sleep-related behaviours associated with rapid eye movement sleep behaviour disorder in Parkinson's disease patients: A case series." *J Clin Pharm Ther* 39 (5): 564–566. doi:10.1111/jcpt.12179.

Chaves, Y. C., K. Genaro, J. A. Crippa, J. M. da Cunha, and J. M. Zanoveli. 2021. "Cannabidiol induces antidepressant and anxiolytic-like effects in experimental type-1 diabetic animals by multiple sites of action." *Metab Brain Dis* 36 (4): 639–652. doi:10.1007/s11011-020-00667-3.

Chaves, Y. C., K. Genaro, C. A. Stern, G. de Oliveira Guaita, J. A. de Souza Crippa, J. M. da Cunha, and J. M. Zanoveli. 2020. "Two-weeks treatment with cannabidiol improves biophysical and behavioral deficits associated with experimental type-1 diabetes." *Neurosci Lett* 729: 135020. doi:10.1016/j.neulet.2020.135020.

Chesher, G. B., C. J. Dahl, M. Everingham, D. M. Jackson, H. Marchant-Williams, and G. A. Starmer. 1973. "The effect of cannabinoids on intestinal motility and their antinociceptive effect in mice." *Br J Pharmacol* 49 (4): 588–594. doi:10.1111/j.1476-5381.1973.tb08534.x.

Chesher, G. B., and D. M. Jackson. 1985. "The quasi-morphine withdrawal syndrome: Effect of cannabinol, cannabidiol and tetrahydrocannabinol." *Pharmacol Biochem Behav* 23 (1): 13–15. doi:10.1016/0091-3057(85)90122-4.

Chesney, E., D. Oliver, A. Green, S. Sovi, J. Wilson, A. Englund, T. P. Freeman, and P. McGuire. 2020. "Adverse effects of cannabidiol: A systematic review and meta-analysis of randomized clinical trials." *Neuropsychopharmacology*. 45 (11): 1799–1806. doi:10.1038/s41386-020-0667-2.

Chesney E, D. Oliver, P. McGuire. 2021 "Cannabidiol (CBD as a novel treatment in the early phases of psychosis." *Psychopharmacology*, online ahead of print. doi: 10.1007/s00213-021-05905-9.

Cipriani, A., T. A. Furukawa, G. Salanti, A. Chaimani, L. Z. Atkinson, Y. Ogawa, S. Leucht, H. G. Ruhe, E. H. Turner, J. P. T. Higgins, M. Egger, N. Takeshima, Y. Hayasaka, H. Imai, K. Shinohara, A. Tajika, J. P. A. Ioannidis, and J. R. Geddes. 2018. "Comparative efficacy and acceptability of 21 antidepressant drugs for the acute treatment of adults with major depressive disorder: A systematic review and network meta-analysis." *Focus (Am Psychiatr Publ)* 16 (4): 420–429. doi:10.1176/appi.focus.16407.

Cluny, N. L., R. J. Naylor, B. A. Whittle, and F. A. Javid. 2008. "The effects of cannabidiol and tetrahydrocannabinol on motion-induced emesis in *Suncus murinus*." *Basic Clin Pharmacol Toxicol* 103 (2): 150–156. doi:10.1111/j.1742-7843.2008.00253.x.

Colasanti, B. K., C. Lindamood III, and C. R. Craig. 1982. "Effects of marihuana cannabinoids on seizure activity in cobalt-epileptic rats." *Pharmacol Biochem Behav* 16 (4): 573–578. doi:10.1016/0091-3057(82)90418-x.

Coles, M., G. Watt, F. Kreilaus, and T. Karl. 2020. "Medium-dose chronic cannabidiol treatment reverses object recognition memory deficits of APP Swe /PS1DeltaE9 transgenic female mice." *Front Pharmacol* 11: 587604. doi:10.3389/fphar.2020.587604.

Collin, C., E. Ehler, G. Waberzinek, Z. Alsindi, P. Davies, K. Powell, W. Notcutt, C. O'Leary, S. Ratcliffe, I. Novakova, O. Zapletalova, J. Pikova, and Z. Ambler. 2010. "A double-blind, randomized, placebo-controlled, parallel-group study of *Sativex*, in subjects with symptoms of spasticity due to multiple sclerosis." *Neurol Res* 32 (5): 451–459. doi:10.1179/016164109X12590518685660.

Comelli, F., G. Giagnoni, I. Bettoni, M. Colleoni, and B. Costa. 2008. "Antihyperalgesic effect of a *Cannabis sativa* extract in a rat model of neuropathic pain: Mechanisms involved." *Phytother Res* 22 (8): 1017–1024. doi:10.1002/ptr.2401.

Consroe, P., E. A. Carlini, A. P. Zwicker, and L. A. Lacerda. 1979. "Interaction of cannabidiol and alcohol in humans." *Psychopharmacology (Berl)* 66 (1): 45–50. doi:10.1007/bf00431988.

Consroe, P., J. Laguna, J. Allender, S. Snider, L. Stern, R. Sandyk, K. Kennedy, and K. Schram. 1991. "Controlled clinical trial of cannabidiol in Huntington's disease." *Pharmacol Biochem Behav* 40 (3): 701–708. doi:10.1016/0091-3057(91)90386-g.

Cordova, T., D. Ayalon, N. Lander, R. Mechoulam, I. Nir, M Puder, and H. R. Lindner. 1980. "The ovulation blocking effect of cannabinoids:

Structure-activity relationships." *Psychoneuroendocrinology* 5: 53–62. doi:10.1016/0306-4530(80)90009-8.

Corroon, J., and J. A. Phillips. 2018. "A cross-sectional study of cannabidiol users." *Cannabis Cannabinoid Res* 3 (1):152–161. doi:10.1089/can.2018.0006.

Costa, B., M. Colleoni, S. Conti, D. Parolaro, C. Franke, A. E. Trovato, and G. Giagnoni. 2004. "Oral anti-inflammatory activity of cannabidiol, a non-psychoactive constituent of cannabis, in acute carrageenan-induced inflammation in the rat paw." *Naunyn Schmiedebergs Arch Pharmacol* 369 (3): 294–299. doi:10.1007/s00210-004-0871-3.

Costa, B., G. Giagnoni, C. Franke, A. E. Trovato, and M. Colleoni. 2004. "Vanilloid TRPV1 receptor mediates the antihyperalgesic effect of the non-psychoactive cannabinoid, cannabidiol, in a rat model of acute inflammation." *Br J Pharmacol* 143 (2): 247–250. doi:10.1038/sj.bjp.0705920.

Costa, B., A. E. Trovato, F. Comelli, G. Giagnoni, and M. Colleoni. 2007. "The non-psychoactive cannabis constituent cannabidiol is an orally effective therapeutic agent in rat chronic inflammatory and neuropathic pain." *Eur J Pharmacol* 556 (1–3): 75–83. doi:10.1016/j.ejphar.2006.11.006.

Costiniuk, C. T., and M. A. Jenabian. 2020. "Acute inflammation and pathogenesis of SARS-CoV-2 infection: Cannabidiol as a potential anti-inflammatory treatment?" *Cytokine Growth Factor Rev.* 53: 63–65. doi:10.1016/j.cytogfr.2020.05.008.

Costiniuk, C. T., Z. Saneei, J. P. Routy, S. Margolese, E. Mandarino, J. Singer, B. Lebouche, J. Cox, J. Szabo, M. J. Brouillette, M. B. Klein, N. Chomont, and M. A. Jenabian. 2019. "Oral cannabinoids in people living with HIV on effective antiretroviral therapy: CTN PT028-study protocol for a pilot randomised trial to assess safety, tolerability and effect on immune activation." *BMJ Open* 9 (1): e024793. doi:10.1136/bmjopen-2018-024793.

Couch, D. G., H. Cook, C. Ortori, D. Barrett, J. N. Lund, and S. E. O'Sullivan. 2019. "Palmitoylethanolamide and cannabidiol prevent inflammation-induced hyperpermeability of the human gut in vitro and in vivo: A randomized, placebo-controlled, double-blind controlled trial." *Inflamm Bowel Dis* 25 (6): 1006–1018. doi:10.1093/ibd/izz017.

Couch, D. G., C. Tasker, E. Theophilidou, J. N. Lund, and S. E. O'Sullivan. 2017. "Cannabidiol and palmitoylethanolamide are anti-inflammatory in the acutely inflamed human colon." *Clin Sci (Lond)* 131 (21): 2611–2626. doi:10.1042/CS20171288.

Cousens, K., and A. DiMascio. 1973. "(−) Delta 9 THC as an hypnotic. An experimental study of three dose levels." *Psychopharmacologia* 33 (4): 355–464. doi:10.1007/bf00437513.

Crawley, J., and F. K. Goodwin. 1980. "Preliminary report of a simple animal behavior model for the anxiolytic effects of benzodiazepines." *Pharmacol Biochem Behav* 13 (2): 167–170. doi:10.1016/0091-3057(80)90067-2.

Crippa, J. A., G. N. Derenusson, T. B. Ferrari, L. Wichert-Ana, F. L. Duran, R. Martin-Santos, M. V. Simoes, S. Bhattacharyya, P. Fusar-Poli, Z. Atakan, A. Santos Filho, M. C. Freitas-Ferrari, P. K. McGuire, A. W. Zuardi, G. F. Busatto, and J. E. Hallak. 2011. "Neural basis of anxiolytic effects of cannabidiol (CBD) in generalized social anxiety disorder: A preliminary report." *J Psychopharmacol* 25 (1): 121–130. doi:10.1177/0269881110379283.

Crippa, J. A., J. E. Hallak, J. P. Machado-de-Sousa, R. H. Queiroz, M. Bergamaschi, M. H. Chagas, and A. W. Zuardi. 2013. "Cannabidiol for the treatment of cannabis withdrawal syndrome: A case report." *J Clin Pharm Ther* 38 (2): 162–164. doi:10.1111/jcpt.12018.

Crippa, J. A. S., A. W. Zuardi, F. S. Guimarães, A. C. Campos, F. de Lima Osório, S. R. Loureiro, R. G. Dos Santos, J. D. S. Souza, J. M. Ushirohira, J. C. Pacheco, R. R. Ferreira, K. C. Mancini Costa, D. S. Scomparin, F. F. Scarante, I. Pires-Dos-Santos, R. Mechoulam, F. Kapczinski, B. A. L. Fonseca, D. L. A. Esposito, K. Pereira-Lima, S. Sen, M. H. Andraus, and J. E. C. Hallak. 2021. "Efficacy and safety of cannabidiol plus standard care vs standard care alone for the treatment of emotional exhaustion and burnout among frontline health care workers during the COVID-19 pandemic: A randomized clinical trial." *JAMA Netw Open* 4 (8): e2120603. doi:10.1001/jamanetworkopen.2021.20603.

Crippa, J. A., A. W. Zuardi, G. E. Garrido, L. Wichert-Ana, R. Guarnieri, L. Ferrari, P. M. Azevedo-Marques, J. E. Hallak, P. K. McGuire, and G. Filho Busatto. 2004. "Effects of cannabidiol (CBD) on regional cerebral blood flow." *Neuropsychopharmacology* 29 (2): 417–426. doi:10.1038/sj.npp.1300340.

Crippa, J. A. S., A. W. Zuardi, J. E. C. Hallak, S. A. Miyazawa, S. A. Bernardo, C. M. Donaduzzi, S. Guzzi, W. A. J. Favreto, A. Campos, M. E. C. Queiroz, F. S. Guimarães, P. M. da Rosa Zimmermann, L. M. Rechia, V. Jose Tondo Filho, and L Brum Junior. 2020. "Oral cannabidiol does not convert to delta-8-THC or delta-9-THC in humans: A pharmacokinetic study in health subjects." *Cannabis Cannabinoid Res* 5 (5): 89–98. doi:10.1089/can.2019.0024.

Crippa, J. A., A. W. Zuardi, R. Martin-Santos, S. Bhattacharyya, Z. Atakan, P. McGuire, and P. Fusar-Poli. 2009. "Cannabis and anxiety: A critical review of the evidence." *Hum Psychopharmacol* 24 (7): 515–523. doi:10.1002/hup.1048.

Criscuolo, E., M. DeSciscio, F. Fezza, and M. Marrarrone. 2020. "In silico and in vitro analysis of major cannabis-derived compounds as fatty acid amide hydrolase inhibitors." *Molecules (Basel, Switzerland)* 26 (1): 48. doi:10.3390/molecules26010048.

Crockett, J., D. Critchley, B. Tayo, J. Berwaerts, and G. Morrison. 2020. "A phase 1, randomized, pharmacokinetic trial of the effect of different meal compositions, whole milk, and alcohol on cannabidiol exposure and safety in healthy subjects." *Epilepsia* 61 (2): 267–277. doi:10.1111/epi.16419.

Cunetti, L., L. Manzo, R. Peyraube, J. Arnaiz, L. Curi, and S. Orihuela. 2018. "Chronic pain treatment with cannabidiol in kidney transplant patients in Uruguay." *Transplant Proc* 50 (2): 461–464. doi:10.1016/j.transproceed.2017.12.042.

Cunha, J. M., E. A. Carlini, A. E. Pereira, O. L. Ramos, C. Pimentel, R. Gagliardi, W. L. Sanvito, N. Lander, and R. Mechoulam. 1980. "Chronic administration of cannabidiol to healthy volunteers and epileptic patients." *Pharmacology* 21 (3): 175–185. doi:10.1159/000137430.

Cuttler, C., E. M. LaFrance, and R. M. Craft. 2020. "A large-scale naturalistic examination of the acute effects of cannabis on pain." *Cannabis Cannabinoid Res*. Online ahead of print. doi:10.1089/can.2020.0068.

Cuttler, C., E. M. LaFrance, and A. Stueber. 2021. "Acute effects of high-potency cannabis flower and cannabis concentrates on everyday life memory and decision making." *Scientific Reports* 11:13784. https://doi.org/10.1038/s41598-021-93198-5

D'Souza, D. C., E. Perry, L. MacDougall, Y. Ammerman, T. Cooper, Y. T. Wu, G. Braley, R. Gueorguieva, and J. H. Krystal. 2004. "The psychotomimetic effects of intravenous delta-9-tetrahydrocannabinol in healthy individuals: Implications for psychosis." *Neuropsychopharmacology* 29 (8): 1558–1572. doi:10.1038/sj.npp.1300496.

Dall'Stella, P. B., M. F. L. Docema, M. V. C. Maldaun, O. Feher, and C. L. P. Lancellotti. 2019. "Case report: Clinical outcome and image response of two patients with secondary high-grade glioma treated with chemoradiation, PCV, and cannabidiol." *Frontiers in Oncology* 8: 643. doi:10.3389/fonc.2018.00643.

Dalton, W. S., R. Martz, L. Lemberger, B. E. Rodda, and R. B. Forney. 1976. "Influence of cannabidiol on delta-9-tetrahydrocannabinol effects." *Clin Pharmacol Ther* 19 (3): 300–309. doi:10.1002/cpt1976193300.

Das, R. K., S. K. Kamboj, M. Ramadas, K. Yogan, V. Gupta, E. Redman, H. V. Curran, and C. J. Morgan. 2013. "Cannabidiol enhances consolidation of explicit fear extinction in humans." *Psychopharmacology (Berl)* 226 (4): 781–792. doi:10.1007/s00213-012-2955-y.

Davies, C., and S. Bhattacharyya. 2019. "Cannabidiol as a potential treatment for psychosis." *Ther Adv Psychopharmacol* 9: 2045125319881916. doi:10.1177/2045125319881916.

de Almeida, C. M. O., M. M. C. Brito, N. B. Bosaipo, A. V. Pimentel, V. Tumas, A. W. Zuardi, J. A. S. Crippa, J. E. C. Hallak, and A. L. Eckeli. 2021. "Cannabidiol for rapid eye movement sleep behavior disorder." *Mov Disord.* 36 (7): 1711–1715. doi:10.1002/mds.28577.

de Faria, S. M., D. de Morais Fabricio, V. Tumas, P. C. Castro, M. A. Ponti, J. E. Hallak, A. W. Zuardi, J. A. S. Crippa, and M. H. N. Chagas. 2020. "Effects of acute cannabidiol administration on anxiety and tremors induced by a simulated public speaking test in patients with Parkinson's disease." *J Psychopharmacol* 34 (2): 189–196. doi:10.1177/0269881119895536.

De Filippis, D., G. Esposito, C. Cirillo, M. Cipriano, B. Y. De Winter, C. Scuderi, G. Sarnelli, R. Cuomo, L. Steardo, J. G. De Man, and T. Iuvone. 2011. "Cannabidiol reduces intestinal inflammation through the control of neuroimmune axis." *PLoS One* 6 (12): e28159. doi:10.1371/journal.pone.0028159.

De Gregorio, D., R. J. McLaughlin, L. Posa, R. Ochoa-Sanchez, J. Enns, M. Lopez-Canul, M. Aboud, S. Maione, S. Comai, and G. Gobbi. 2019. "Cannabidiol modulates serotonergic transmission and reverses both allodynia and anxiety-like behavior in a model of neuropathic pain." *Pain* 160 (1): 136–150. doi:10.1097/j.pain.0000000000001386.

De Petrocellis, L. P. Orlando, A. S. Moriello, G. Aviello, C. Stott, A. A. Izzo, and V. Di Marzo. 2012. "Cannabinoid actions at TRPV channels:

Effects on TRPV3 and TRPV4 and their potential relevance to gastrointestinal inflammation." *Acta Physiol (Oxf)* 204 (2): 255–266. doi:10.1111/j.1748-1716.2011.02338.x.

De Ternay, J., M. Naassila, M. Nourredine, A. Louvet, F. Bailly, G. Sescousse, P. Maurage, O. Cottencin, P. M. Carrieri, and B. Rolland. 2019. "Therapeutic prospects of cannabidiol for alcohol use disorder and alcohol-related damages on the liver and the brain." *Front Pharmacol* 10: 627. doi:10.3389/fphar.2019.00627.

DeJesus, E., B. M. Rodwick, D. Bowers, C. J. Cohen, and D. Pearce. 2007. "Use of dronabinol improves appetite and reverses weight loss in HIV/AIDS-INFECTED PATIEnts." *J Int Assoc Physicians AIDS Care (Chic)* 6 (2): 95–100. doi:10.1177/1545109707300157.

Devane, W. A., L. Hanus, A. Breuer, R. G. Pertwee, L. A. Stevenson, G. Griffin, D. Gibson, A. Mandelbaum, A. Etinger, and R. Mechoulam. 1992. "Isolation and structure of a brain constituent that binds to the cannabinoid receptor." *Science* 258 (5090): 1946–1949. doi:10.1126/science.1470919.

Devinsky, O., M. R. Cilio, H. Cross, J. Fernandez-Ruiz, J. French, C. Hill, R. Katz, V. Di Marzo, D. Jutras-Aswad, W. G. Notcutt, J. Martinez-Orgado, P. J. Robson, B. G. Rohrback, E. Thiele, B. Whalley, and D. Friedman. 2014. "Cannabidiol: Pharmacology and potential therapeutic role in epilepsy and other neuropsychiatric disorders." *Epilepsia* 55 (6): 791–802. doi:10.1111/epi.12631.

Devinsky, O., J. H. Cross, L. Laux, E. Marsh, I. Miller, R. Nabbout, I. E. Scheffer, E. A. Thiele, S. Wright, and Group Cannabidiol in Dravet Syndrome Study. 2017. "Trial of cannabidiol for drug-resistant seizures in the Dravet syndrome." *N Engl J Med* 376 (21): 2011–2020. doi:10.1056/NEJMoa1611618.

Devinsky, O., E. Marsh, D. Friedman, E. Thiele, L. Laux, J. Sullivan, I. Miller, R. Flamini, A. Wilfong, F. Filloux, M. Wong, N. Tilton, P. Bruno, J. Bluvstein, J. Hedlund, R. Kamens, J. Maclean, S. Nangia, N. S. Singhal, C. A. Wilson, A. Patel, and M. R. Cilio. 2016. "Cannabidiol in patients with treatment-resistant epilepsy: An open-label interventional trial." *Lancet Neurol* 15 (3): 270–278. doi:10.1016/S1474-4422(15)00379-8.

Devinsky, O., A. D. Patel, J. H. Cross, V. Villanueva, E. C. Wirrell, M. Privitera, S. M. Greenwood, C. Roberts, D. Checketts, K. E. VanLandingham, and S. M. Zuberi. 2018. "Effect of cannabidiol on drop seizures in the

Lennox-Gastaut syndrome." *N Engl J Med* 378 (20): 1888–1897. doi:10.1056/NEJMoa1714631.

Devinsky, O., A. D. Patel, J. H. Cross, V. Villanueva, E. C. Wirrell, M. Privitera, S. M. Greenwood, C. Roberts, D. Checketts, K. E. VanLandingham, S. M. Zuberi, and Gwpcare Study Group. 2018. "Effect of cannabidiol on drop seizures in the Lennox-Gastaut syndrome." *N Engl J Med* 378 (20): 1888–1897. doi:10.1056/NEJMoa1714631.

Devinsky, O., A. D. Patel, E. A. Thiele, M. H. Wong, R. Appleton, C. L. Harden, S. Greenwood, G. Morrison, K. Sommerville, and Gwpcare Part A Study Group. 2018. "Randomized, dose-ranging safety trial of cannabidiol in Dravet syndrome." *Neurology* 90 (14): e1204-e1211. doi:10.1212/WNL.0000000000005254.

De Vita, M. J., S. A. Maisto, C. E. Gilmour, L. McGuire, E. Tarvin, and D. Moskal. 2021. "The effects of cannabidiol and analgesic expectancies on experimental pain reactivity in healthy adults: A balanced placebo design trial." *Exp Clin Psychopharmacol* doi:10.1037/pha0000465.

DeVuono, M. V., O. La Caprara, G. N. Petrie, C. L. Limebeer, E. M. Rock, M. N. Hill, and L. A. Parker. 2020. "Cannabidiol Interferes with establishment of Δ 9—tetrahydrocannabinol-induced nausea through a 5-HT 1A mechanism." *Cannabis and Cannabinoid Research*. Online ahead of print. doi:10.1089/can.2020.0083.

DeVuono, M. V., and L. A. Parker. 2020. "Cannabinoid hyperemesis syndrome: A review of potential mechanisms." *Cannabis and Cannabinoid Research*. 5 (2): 132–144. doi:10.1089/can.2019.0059.

Dezieck, L., Z. Hafez, A. Conicella, E. Blohm, M. J. O'Connor, E. S. Schwarz, and M.l E. Mullins. 2017. "Resolution of cannabis hyperemesis syndrome with topical capsaicin in the emergency department: A case series." *Clinical Toxicology* 55 (8): 908–913. doi:10.1080/15563650.2017.1324166.

Di Forti, M., C. Morgan, P. Dazzan, C. Pariante, V. Mondelli, T. R. Marques, R. Handley, S. Luzi, M. Russo, A. Paparelli, A. Butt, S. A. Stilo, B. Wiffen, J. Powell, and R. M. Murray. 2009. "High-potency cannabis and the risk of psychosis." *Br J Psychiatry* 195 (6): 488–491. doi:10.1192/bjp.bp.109.064220.

di Giacomo, V., A. Chiavaroli, G. Orlando, A. Cataldi, M. Rapino, V. Di Valerio, S. Leone, L. Brunetti, L. Menghini, L. Recinella, and C. Ferrante. 2020. "Neuroprotective and neuromodulatory effects induced by cannabidiol and cannabigerol in rat hypo-E22 cells and isolated hypothalamus." *Antioxidants (Basel)* 9 (1): 71. doi:10.3390/antiox9010071.

Do Monte, F. H., R. R. Souza, R. M. Bitencourt, J. A. Kroon, and R. N. Takahashi. 2013. "Infusion of cannabidiol into infralimbic cortex facilitates fear extinction via CB1 receptors." *Behav Brain Res* 250: 23–27. doi:10.1016/j.bbr.2013.04.045.

Duran, M., E. Perez, S. Abanades, X. Vidal, C. Saura, M. Majem, E. Arriola, M. Rabanal, A. Pastor, M. Farre, N. Rams, J. R. Laporte, and D. Capella. 2010. "Preliminary efficacy and safety of an oromucosal standardized cannabis extract in chemotherapy-induced nausea and vomiting." *Br J Clin Pharmacol* 70 (5): 656–663. doi:10.1111/j.1365-2125.2010.03743.x.

Durst, R., H. Danenberg, R. Gallily, R. Mechoulam, K. Meir, E. Grad, R. Beeri, T. Pugatsch, E. Tarsish, and C. Lotan. 2007. "Cannabidiol, a nonpsychoactive cannabis constituent, protects against myocardial ischemic reperfusion injury." *Am J Physiol Heart Circ Physiol* 293 (6): H3602–607. doi:10.1152/ajpheart.00098.2007.

Edery, H., Y. Grunfeld, G. Porath, Z. Ben-Zvi, A. Shani, and R. Mechoulam. 1972. "Structure-activity relationships in the tetrahydrocannabinol series. Modifications on the aromatic ring and in the side-chain." *Arzneimittelforschung* 22 (11): 1995–2003.

El-Alfy, A. T., K. Ivey, K. Robinson, S. Ahmed, M. Radwan, D. Slade, I. Khan, M. ElSohly, and S. Ross. 2010. "Antidepressant-like effect of delta9-tetrahydrocannabinol and other cannabinoids isolated from *Cannabis sativa* L." *Pharmacol Biochem Behav* 95 (4): 434–442. doi:10.1016/j.pbb.2010.03.004.

Elms, L., S. Shannon, S. Hughes, and N. Lewis. 2019. "Cannabidiol in the treatment of post-traumatic stress disorder: A case series." *J Altern Complement Med* 25 (4): 392–397. doi:10.1089/acm.2018.0437.

ElSohly, M. A., S. Chandra, M. Radwan, C. G. Majumdar, and J. C. Church. 2021. "A comprehensive review of cannabis potency in the United States in the last decade." *Biol Psychiatry Cogn Neurosci Neuroimaging*. 6 (6): 603–606. doi:10.1016/j.bpsc.2020.12.016.

ElSohly, M. A., Z. Mehmedic, S. Foster, C. Gon, S. Chandra, and J. C. Church. 2016. "Changes in cannabis potency over the last 2 decades (1995–2014): Analysis of current data in the United States." *Biol Psychiatry* 79 (7): 613–619. doi:10.1016/j.biopsych.2016.01.004.

ElSohly, M., and W. Gul. 2014. "Constituents of *Cannabis sativa*." In *Handbook of Cannabis*, edited by Roger G. Pertwee, 3–22. Oxford: Oxford University Press.

Englund, A., T. P. Freeman, R. M. Murray, and P. McGuire. 2017. "Can we make cannabis safer?" *Lancet Psychiatry* 4 (8): 643–648. doi:10.1016/S2215-0366(17)30075-5.

Englund, A., P. D. Morrison, J. Nottage, D. Hague, F. Kane, S. Bonaccorso, J. M. Stone, A. Reichenberg, R. Brenneisen, D. Holt, A. Feilding, L. Walker, R. M. Murray, and S. Kapur. 2013. "Cannabidiol inhibits THC-elicited paranoid symptoms and hippocampal-dependent memory impairment." *J Psychopharmacol* 27 (1): 19–27. doi:10.1177/0269881112460109.

Eskander, J. P., J. Spall, A. Spall, R. V. Shah, and A. D. Kaye. 2020. "Cannabidiol (CBD) as a treatment of acute and chronic back pain: A case series and literature review." *J Opioid Manag* 16 (3): 215–218. doi:10.5055/jom.2020.0570.

Evans, M. A., R. Martz, B. E. Rodda, L. Lemberger, and R. B. Forney. 1976. "Effects of marihuana-dextroamphetamine combination." *Clin Pharmacol Ther* 20 (3): 350–358. doi:10.1002/cpt1976203350.

Fallon, M. T., E. Albert Lux, R. McQuade, S. Rossetti, R. Sanchez, W. Sun, S. Wright, A. H. Lichtman, and E. Kornyeyeva. 2017. "Sativex oromucosal spray as adjunctive therapy in advanced cancer patients with chronic pain unalleviated by optimized opioid therapy: Two double-blind, randomized, placebo-controlled phase 3 studies." *Br J Pain* 11 (3): 119–133. doi:10.1177/2049463717710042.

Farrimond, J. A., B. J. Whalley, and C. M. Williams. 2012. "Cannabinol and cannabidiol exert opposing effects on rat feeding patterns." *Psychopharmacology (Berl)* 223 (1): 117–129. doi:10.1007/s00213-012-2697-x.

Feng, Y., F. Chen, T. Yin, Q. Xia, Y. Liu, G. Huang, J. Zhang, R. Oyen, and Y. Ni. 2015. "Pharmacologic effects of cannabidiol on acute reperfused myocardial infarction in rabbits: Evaluated with 3.0T cardiac magnetic resonance imaging and histopathology." *J Cardiovasc Pharmacol* 66 (4): 354–363. doi:10.1097/FJC.0000000000000287.

Fernández-Ruiz, J., E. de Lago, M. Gómez-Ruiz, C. García, O. Sagredo, and M. García-Arencibia. 2014. "Neurodegenerative disorders other than multiple sclerosis." In *Handbook of Cannabis*, edited by Roger G. Pertwee, 505–525. Oxford: Oxford University Press.

Ferre, L., A. Nuara, G. Pavan, M. Radaelli, L. Moiola, M. Rodegher, B. Colombo, I. J. Keller Sarmiento, V. Martinelli, L. Leocani, F. Martinelli Boneschi, G. Comi, and F. Esposito. 2016. "Efficacy and safety of nabiximols (Sativex) on multiple sclerosis spasticity in a real-life Italian monocentric study." *Neurol Sci* 37 (2): 235–242. doi:10.1007/s10072-015-2392-x.

Finn, D. P., S. R. Beckett, C. H. Roe, A. Madjd, K. C. Fone, D. A. Kendall, C. A. Marsden, and V. Chapman. 2004. "Effects of coadministration of cannabinoids and morphine on nociceptive behaviour, brain monoamines and HPA axis activity in a rat model of persistent pain." *Eur J Neurosci* 19 (3): 678–686. doi:10.1111/j.0953-816x.2004.03177.x.

Flachenecker, P., T. Henze, and U. K. Zettl. 2014. "Nabiximols (THC/CBD oromucosal spray, Sativex in clinical practice—results of a multicenter, non-interventional study (MOVE 2) in patients with multiple sclerosis spasticity." *Eur Neurol* 71 (5–6): 271–279. doi:10.1159/000357427.

Florensa-Zanuy, E., E. Garro-Martínez, A. Adell, E. Castro, Á Díaz, Á Pazos, K. S. Mac-Dowell, D. Martín-Hernández, and F. Pilar-Cuéllar. 2021. "Cannabidiol antidepressant-like effect in the lipopolysaccharide model in mice: Modulation of inflammatory pathways." *Biochem Pharmacol* 185: 114433. doi:10.1016/j.bcp.2021.114433.

Fogaca, M. V., F. M. Reis, A. C. Campos, and F. S. Guimaraes. 2014. "Effects of intra-prelimbic prefrontal cortex injection of cannabidiol on anxiety-like behavior: Involvement of 5HT1A receptors and previous stressful experience." *Eur Neuropsychopharmacol* 24 (3): 410–419. doi:10.1016/j.euroneuro.2013.10.012.

Forget, B., K. M. Coen, and B. LeFoll. 2009. "Inhibition of fatty acid amide hydrolase reduces reinstatement of nicotine seeking but not break point for nicotine self-administration—comparison with CB1 receptor blockade." *Psychopharmacology* 205 (4): 613–624. doi:10.1007/s00213-009-1569-5.

Formukong, E. A., A. T. Evans, and F. J. Evans. 1988. "Analgesic and antiinflammatory activity of constituents of *Cannabis sativa L*." *Inflammation* 12 (4): 361–371. doi:10.1007/bf00915771.

Foss, J. D., D. J. Farkas, L. M. Huynh, W. A. Kinney, D. E. Brenneman, and S. J. Ward. 2021. "Behavioural and pharmacological effects of cannabidiol (CBD) and the cannabidiol analogue KLS-13019 in mouse models of pain and reinforcement." *Br J Pharmacol* 178 (15): 3067–3078. doi:10.1111/bph.15486.

Fouad, A. A., A. S. Al-Mulhim, and W. Gomaa. 2013. "Protective effect of cannabidiol against cadmium hepatotoxicity in rats." *J Trace Elem Med Biol* 27 (4): 355–363. doi:10.1016/j.jtemb.2013.07.001.

Freeman, A. M., K. Petrilli, R. Lees, C. Hindocha, C. Mokrysz, H. V. Curran, R. Saunders, and T. P. Freeman. 2019. "How does cannabidiol (CBD) influence the acute effects of delta-9-tetrahydrocannabinol (THC) in humans? A systematic review." *Neurosci Biobehav Rev* 107: 696–712. doi:10.1016/j.neubiorev.2019.09.036.

Freeman, T. P., R. A. Pope, M. B. Wall, J. A. Bisby, M. Luijten, C. Hindocha, C. Mokrysz, W. Lawn, A. Moss, M. A. P. Bloomfield, C. J. A. Morgan, D. J. Nutt, and H. V. Curran. 2018. "Cannabis dampens the effects of music in brain regions sensitive to reward and emotion." *Int J Neuropsychopharmacol* 21 (1): 21–32. doi:10.1093/ijnp/pyx082.

French, E. D., K. Dillon, and X. Wu. 1997. "Cannabinoids excite dopamine neurons in the ventral tegmentum and substantia nigra." *Neuroreport* 8 (3): 649–652. doi:10.1097/00001756-199702100-00014.

Fride, E., C. Feigin, D. E. Ponde, A. Breuer, L. Hanus, N. Arshavsky, and R. Mechoulam. 2004. "(+)-Cannabidiol analogues which bind cannabinoid receptors but exert peripheral activity only." *Eur J Pharmacol* 506 (2): 179–188. doi:10.1016/j.ejphar.2004.10.049.

Friedman, D., J. A. French, and M. Maccarrone. 2019. "Safety, efficacy, and mechanisms of action of cannabinoids in neurological disorders." *Lancet Neurol* 18 (5): 504–512. doi:10.1016/S1474-4422(19)30032-8.

Fu, B., X. Xu, and H. Wei. 2020. "Why tocilizumab could be an effective treatment for severe COVID-19?" *Journal of Translational Medicine* 18 (1): 164. doi:10.1186/s12967-020-02339-3.

Fusar-Poli, P., P. Allen, S. Bhattacharyya, J. A. Crippa, A. Mechelli, S. Borgwardt, R. Martin-Santos, M. L. Seal, C. O'Carrol, Z. Atakan, A. W. Zuardi, and P. McGuire. 2010. "Modulation of effective connectivity during emotional processing by delta 9-tetrahydrocannabinol and cannabidiol." *Int J Neuropsychopharmacol* 13 (4): 421–432. doi:10.1017/S1461145709990617.

Fusar-Poli, P., J. A. Crippa, S. Bhattacharyya, S. J. Borgwardt, P. Allen, R. Martin-Santos, M. Seal, S. A. Surguladze, C. O'Carrol, Z. Atakan, A. W. Zuardi, and P. K. McGuire. 2009. "Distinct effects of {delta}9-tetrahydrocannabinol and cannabidiol on neural activation during emotional processing." *Arch Gen Psychiatry* 66 (1): 95–105. doi:10.1001/archgenpsychiatry.2008.519.

Galaj, E., G. H. Bi, H. J. Yang, and Z. X. Xi. 2020. "Cannabidiol attenuates the rewarding effects of cocaine in rats by CB2, 5-HT1A and TRPV1 receptor mechanisms." *Neuropharmacology* 167: 107740. doi:10.1016/j.neuropharm.2019.107740.

Gallily, R., T. Even-Chena, G. Katzavian, D. Lehmann, A. Dagan, and R. Mechoulam. 2003. "Gamma-irradiation enhances apoptosis induced by cannabidiol, a non-psychotropic cannabinoid, in cultured HL-60 myeloblastic leukemia cells." *Leuk Lymphoma* 44 (10): 1767–173. doi:10.1080/1042819031000103917.

Gamble, L. J., J. M. Boesch, C. W. Frye, W. S. Schwark, S. Mann, L. Wolfe, H. Brown, E. S. Berthelsen, and J. J. Wakshlag. 2018. "Pharmacokinetics, safety, and clinical efficacy of cannabidiol treatment in osteoarthritic dogs." *Front Vet Sci* 5: 165. doi:10.3389/fvets.2018.00165.

Gaoni, Y., and R. Mechoulam. 1964. "Hashish. III. Isolation, structure, and partial synthesis of an active constituent of hashish." *Journal of the American Chemical Society* 86: 1646–1647.

Gaoni, Y., and R. Mechoulam. 1966a. "The isomerization of cannabidiol to tetrahydrocannabinols." *Tetrahedron* 22: 1481–1488.

Gaoni, Y., and R. Mechoulam. 1966b. "Concerning the isomerization of delta1 to delta6-tetrahydrocannabinol." *J Amer Chem Soc* 88: 5673–5675. doi:10.1089/can.2019.0024.

Gaston, T. E., E. M. Bebin, G. R. Cutter, S. B. Ampah, Y. Liu, L. P. Grayson, J. P. Szaflarski, and UAB CBD Program. 2019. "Drug-drug interactions with cannabidiol (CBD) appear to have no effect on treatment response in an open-label expanded access program." *Epilepsy Behav* 98 (Pt. A): 201–206. doi:10.1016/j.yebeh.2019.07.008.

Gaston, T. E., E. M. Bebin, G. R. Cutter, Y. Liu, J. P. Szaflarski, and UAB CBD Program. 2017. "Interactions between cannabidiol and commonly used antiepileptic drugs." *Epilepsia* 58 (9): 1586–1592. doi:10.1111/epi.13852.

Geffrey, A. L., S. F. Pollack, P. L. Bruno, and E. A. Thiele. 2015. "Drug-drug interaction between clobazam and cannabidiol in children with refractory epilepsy." *Epilepsia* 56 (8): 1246–1251. doi:10.1111/epi.13060.

Genaro, K., D. Fabris, A. L. F. Arantes, A. W. Zuardi, J. A. S. Crippa, and W. A. Prado. 2017. "Cannabidiol is a potential therapeutic for the affective-motivational dimension of incision pain in rats." *Front Pharmacol* 8: 391. doi:10.3389/fphar.2017.00391.

Ghabrash, M. F., S. Coronado-Montoya, J. Aoun, A. A. Gagne, F. Mansour, C. Ouellet-Plamondon, A. Trepanier, and D. Jutras-Aswad. 2020. "Cannabidiol for the treatment of psychosis among patients with schizophrenia and other primary psychotic disorders: A systematic review with a risk of bias assessment." *Psychiatry Res* 286: 112890. doi:10.1016/j.psychres.2020.112890.

Giacoppo, S., M. Galuppo, F. Pollastro, G. Grassi, P. Bramanti, and E. Mazzon. 2015. "A new formulation of cannabidiol in cream shows therapeutic effects in a mouse model of experimental autoimmune encephalomyelitis." *Daru* 23: 48. doi:10.1186/s40199-015-0131-8.

Giuffrida, A., F. M. Leweke, C. W. Gerth, D. Schreiber, D. Koethe, J. Faulhaber, J. Klosterkotter, and D. Piomelli. 2004. "Cerebrospinal anandamide levels are elevated in acute schizophrenia and are inversely correlated with psychotic symptoms." *Neuropsychopharmacology* 29 (11): 2108–2114. doi:10.1038/sj.npp.1300558.

Goldstein, R. B., S. M. Smith, S. P. Chou, T. D. Saha, J. Jung, H. Zhang, R. P. Pickering, W. J. Ruan, B. Huang, and B. F. Grant. 2016. "The epidemiology of DSM-5 posttraumatic stress disorder in the United States: Results from the National Epidemiologic Survey on Alcohol and Related Conditions–III." *Soc Psychiatry Psychiatr Epidemiol* 51 (8): 1137–1148. doi:10.1007/s00127-016-1208-5.

Gomes, F. V., E. A. Del Bel, and F. S. Guimaraes. 2013. "Cannabidiol attenuates catalepsy induced by distinct pharmacological mechanisms via 5-HT1A receptor activation in mice." *Prog Neuropsychopharmacol Biol Psychiatry* 46: 43–47. doi:10.1016/j.pnpbp.2013.06.005.

Gomes, F. V., A. C. Issy, F. R. Ferreira, M. P. Viveros, E. A. Del Bel, and F. S. Guimaraes. 2014. "Cannabidiol attenuates sensorimotor gating disruption and molecular changes induced by chronic antagonism of NMDA receptors in mice." *Int J Neuropsychopharmacol* 18 (5): pyu041. doi:10.1093/ijnp/pyu041.

Gomes, F. V., D. G. Reis, F. H. Alves, F. M. Correa, F. S. Guimaraes, and L. B. Resstel. 2012. "Cannabidiol injected into the bed nucleus of the stria terminalis reduces the expression of contextual fear conditioning via 5-HT1A receptors." *J Psychopharmacol* 26 (1): 104–113. doi:10.1177/0269881110389095.

Gomes, F. V., L. B. Resstel, and F. S. Guimaraes. 2011. "The anxiolytic-like effects of cannabidiol injected into the bed nucleus of the stria terminalis are mediated by 5-HT1A receptors." *Psychopharmacology (Berl)* 213 (2–3): 465–473. doi:10.1007/s00213-010-2036-z.

Gonzalez-Cuevas, G., R. Martin-Fardon, T. M. Kerr, D. G. Stouffer, L. H. Parsons, D. C. Hammell, S. L. Banks, A. L. Stinchcomb, and F. Weiss. 2018. "Unique treatment potential of cannabidiol for the prevention of relapse to drug use: Preclinical proof of principle." *Neuropsychopharmacology* 43 (10): 2036–2045. doi:10.1038/s41386-018-0050-8.

Good, P. D., R. M. Greer, G. E. Huggett, and J. R. Hardy. 2020. "An open-label pilot study testing the feasibility of assessing total symptom burden in trials of cannabinoid medications in palliative care." *Journal of Palliative Medicine* 23 (5): 650–655. doi:10.1089/jpm.2019.0540.

Greene, N. Z., J. L. Wiley, Z. Yu, B. H. Clowers, and R. M. Craft. 2018. "Cannabidiol modulation of antinociceptive tolerance to delta(9)-tetrahydrocannabinol." *Psychopharmacology (Berl)* 235 (11): 3289–3302. doi:10.1007/s00213-018-5036-z.

Grill, H. J., and R. Norgren. 1978. "The taste reactivity test. I. Mimetic responses to gustatory stimuli in neurologically normal rats." *Brain Res* 143 (2): 263–279.

Grim, T. W., S. Ghosh, K. L. Hsu, B. F. Cravatt, S. G. Kinsey, and A. H. Lichtman. 2014. "Combined inhibition of FAAH and COX produces enhanced anti-allodynic effects in mouse neuropathic and inflammatory pain models." *Pharmacol Biochem Behav* 124: 405–411. doi:10.1016/j.pbb.2014.07.008.

Grimison, P., A. Mersiades, A. Kirby, N. Lintzeris, R. Morton, P. Haber, I. Olver, A. Walsh, I. McGregor, Y. Cheung, A. Tognela, C. Hahn, K. Briscoe, M. Aghmesheh, P. Fox, E. Abdi, S. Clarke, S. Della-Fiorentina, J. Shannon, C. Gedye, S. Begbie, J. Simes, and M. Stockler. 2020. "Oral THC:CBD cannabis extract for refractory chemotherapy-induced nausea and vomiting: A randomised, placebo-controlled, phase II crossover trial." *Ann Oncol* 31 (11): 1553–1560. doi:10.1016/j.annonc.2020.07.020.

Grotenhermen, F. 2003. "Pharmacokinetics and pharmacodynamics of cannabinoids." *Clin Pharmacokinet* 42 (4): 327–360. doi:10.2165/00003088-200342040-00003.

Grotenhermen, F., E. Russo, and A. W. Zuardi. 2017. "Even high doses of oral cannabidol do not cause THC-like Effects in humans: Comment on Merrick et al. cannabis and cannabinoid research 2016;1(1): 102–112;

DOI:10.1089/can.2015.0004." *Cannabis Cannabinoid Res* 2 (1): 1–4. doi:10.1089/can.2016.0036.

Guimaraes, F. S., T. M. Chiaretti, F. G. Graeff, and A. W. Zuardi. 1990. "Antianxiety effect of cannabidiol in the elevated plus-maze." *Psychopharmacology (Berl)* 100 (4): 558–559.

Gulbransen, G., W. Xu, and B. Arroll. 2020. "Cannabidiol prescription in clinical practice: An audit on the first 400 patients in New Zealand." *BJGP Open* 4 (1): bjgpopen20X101010. doi:10.3399/bjgpopen20X101010.

Gururajan, A., and D. T. Malone. 2016. "Does cannabidiol have a role in the treatment of schizophrenia?" *Schizophr Res* 176 (2–3): 281–290. doi:10.1016/j.schres.2016.06.022.

Guy, G. W., Robson, P. J. 2003. "A phase I, double blind, three-way crossover study to assess the pharmacokinetic profile of cannabis-based medicine extract (CBME) administered sublingually in variant cannabinoid ratios in normal healthy male volunteers (GWPK0215)." *J Cannabis Ther* 3: 121–152. doi:10.1300/J175v03n04_02

Hacke, A. C. M., D. Lima, F. de Costa, K. Deshmukh, N. Li, A. M. Chow, J. A. Marques, R. P. Pereira, and K. Kerman. 2019. "Probing the antioxidant activity of delta(9)-tetrahydrocannabinol and cannabidiol in *Cannabis sativa* extracts." *Analyst* 144 (16): 4952–4961. doi:10.1039/c9an00890j.

Halladay, J. E., J. MacKillop, C. Munn, S. M. Jack, and K. Georgiades. 2020. "Cannabis use as a risk factor for depression, anxiety, and suicidality: Epidemiological associations and implications for nurses." *J Addict Nurs* 31 (2): 92–101. doi:10.1097/jan.0000000000000334.

Hamelink, C., A. Hampson, D. A. Wink, L. E. Eiden, and R. L. Eskay. 2005. "Comparison of cannabidiol, antioxidants, and diuretics in reversing binge ethanol-induced neurotoxicity." *J Pharmacol Exp Ther* 314 (2): 780–788. doi:10.1124/jpet.105.085779.

Hammell, D. C., L. P. Zhang, F. Ma, S. M. Abshire, S. L. McIlwrath, A. L. Stinchcomb, and K. N. Westlund. 2016. "Transdermal cannabidiol reduces inflammation and pain-related behaviours in a rat model of arthritis." *Eur J Pain* 20 (6): 936–948. doi:10.1002/ejp.818.

Hampson, A. J., M. Grimaldi, J. Axelrod, and D. Wink. 1998. "Cannabidiol and (−)delta9-tetrahydrocannabinol are neuroprotective antioxidants." *Proc Natl Acad Sci USA* 95 (14): 8268–8273. doi:10.1073/pnas.95.14.8268.

Hampson, A. J., M. Grimaldi, M. Lolic, D. Wink, R. Rosenthal, and J. Axelrod. 2000. "Neuroprotective antioxidants from marijuana." *Ann NY Acad Sci* 899: 274–282.

Haney, M. 2020. "Perspectives on cannabis research-barriers and recommendations." *JAMA Psychiatry* 77: 994–995. doi:10.1001/jamapsychiatry.2020.103.

Haney, M., E. W. Gunderson, J. Rabkin, C. L. Hart, S. K. Vosburg, S. D. Comer, and R. W. Foltin. 2007. "Dronabinol and marijuana in HIV-positive marijuana smokers. Caloric intake, mood, and sleep." *J Acquir Immune Defic Syndr* 45 (5): 545–554. doi:10.1097/QAI.0b013e31811ed205.

Haney, M., R. J. Malcolm, S. Babalonis, P. A. Nuzzo, Z. D. Cooper, G. Bedi, K. M. Gray, A. McRae-Clark, M. R. Lofwall, S. Sparenborg, and S. L. Walsh. 2016. "Oral cannabidiol does not alter the subjective, reinforcing or cardiovascular effects of smoked cannabis." *Neuropsychopharmacology* 41 (8): 1974–1982. doi:10.1038/npp.2015.367.

Haney, M., J. Rabkin, E. Gunderson, and R. W. Foltin. 2005. "Dronabinol and marijuana in HIV(+) marijuana smokers: Acute effects on caloric intake and mood." *Psychopharmacology (Berl)* 181 (1): 170–178. doi:10.1007/s00213-005-2242-2.

Hanus, L. O., S. Tchilibon, D. E. Ponde, A. Breuer, E. Fride, and R. Mechoulam. 2005. "Enantiomeric cannabidiol derivatives: Synthesis and binding to cannabinoid receptors." *Org Biomol Chem* 3 (6): 1116–1123 . doi:10.1039/b416943c.

Hao, E., P. Mukhopadhyay, Z. Cao, K. Erdelyi, E. Holovac, L. Liaudet, W. S. Lee, G. Hasko, R. Mechoulam, and P. Pacher. 2015. "Cannabidiol protects against doxorubicin-induced cardiomyopathy by modulating mitochondrial function and biogenesis." *Mol Med* 21: 38–45. doi:10.2119/molmed.2014.00261.

Harris, H. M., K. J. Sufka, W. Gul, and M. A. ElSohly. 2016. "Effects of delta-9-tetrahydrocannabinol and cannabidiol on cisplatin-induced neuropathy in mice." *Planta Med* 82 (13): 1169–1172. doi:10.1055/s-0042–106303.

Haupts, M., C. Vila, A. Jonas, K. Witte, and L. Alvarez-Ossorio. 2016. "Influence of previous failed antispasticity therapy on the efficacy and tolerability of THC:CBD oromucosal spray for multiple sclerosis spasticity." *Eur Neurol* 75 (5–6): 236–243. doi:10.1159/000445943.

Hay, G. L., S. J. Baracz, N. A. Everett, J. Roberts, P. A. Costa, J. C. Arnold, I. S. McGregor, and J. L. Cornish. 2018. "Cannabidiol treatment reduces the motivation to self-administer methamphetamine and methamphetamine-primed relapse in rats." *J Psychopharmacol* 32 (12): 1369–1378. doi:10.1177 /0269881118799954.

Hayakawa, K., K. Mishima, K. Irie, M. Hazekawa, S. Mishima, M. Fujioka, K. Orito, N. Egashira, S. Katsurabayashi, K. Takasaki, K. Iwasaki, and M. Fujiwara. 2008. "Cannabidiol prevents a post-ischemic injury progressively induced by cerebral ischemia via a high-mobility group box1-inhibiting mechanism." *Neuropharmacology* 55 (8): 1280–1286. doi:10.1016/j.neuro pharm.2008.06.040.

Hayakawa, K., K. Mishima, M. Nozako, A. Ogata, M. Hazekawa, A. X. Liu, M. Fujioka, K. Abe, N. Hasebe, N. Egashira, K. Iwasaki, and M. Fujiwara. 2007. "Repeated treatment with cannabidiol but not delta9-tetrahydrocannabinol has a neuroprotective effect without the development of tolerance." *Neuropharmacology* 52 (4): 1079–1087. doi:10.1016/j.neuropharm.2006.11.005.

Hegde, V. L., P. S. Nagarkatti, and M. Nagarkatti. 2011. "Role of myeloid-derived suppressor cells in amelioration of experimental autoimmune hepatitis following activation of TRPV1 receptors by cannabidiol." *PLoS One* 6 (4): e18281. doi:10.1371/journal.pone.0018281.

Hen-Shoval, D., S. Amar, L. Shbiro, R. Smoum, C. G. Haj, R. Mechoulam, G. Zalsman, A. Weller, and G. Shoval. 2018. "Acute oral cannabidiolic acid methyl ester reduces depression-like behavior in two genetic animal models of depression." *Behav Brain Res* 351: 1–3. doi:10.1016/j.bbr.2018.05.027.

Herlopian, A., E. J. Hess, J. Barnett, A. L. Geffrey, S. F. Pollack, L. Skirvin, P. Bruno, J. Sourbron, and E. A. Thiele. 2020. "Cannabidiol in treatment of refractory epileptic spasms: An open-label study." *Epilepsy Behav* 106: 106988. doi:10.1016/j.yebeh.2020.106988.

Hickok, J. T., J. A. Roscoe, G. R. Morrow, D. K. King, J. N. Atkins, and T. R. Fitch. 2003. "Nausea and emesis remain significant problems of chemotherapy despite prophylaxis with 5-hydroxytryptamine-3 antiemetics: A University of Rochester James P. Wilmot Cancer Center Community Clinical Oncology Program Study of 360 cancer patients treated in the community." *Cancer* 97 (11): 2880–2886. doi:10.1002/cncr.11408.

Highet, B. H., E. R. Lesser, P. W. Johnson, and J. S. Kaur. 2020. "Tetrahydrocannabinol and cannabidiol use in an outpatient palliative medicine population." *American Journal of Hospice and Palliative Medicine* 37 (8): 589–593. doi:10.1177/1049909119900378.

Hill, A. J., M. S. Mercier, T. D. Hill, S. E. Glyn, N. A. Jones, Y. Yamasaki, T. Futamura, M. Duncan, C. G. Stott, G. J. Stephens, C. M. Williams, and B. J. Whalley. 2012. "Cannabidivarin is anticonvulsant in mouse and rat." *Br J Pharmacol* 167 (8): 1629–1642. doi:10.1111/j.1476-5381.2012.02207.x.

Hill, K. P. 2020. "Cannabinoids and the coronavirus." *Cannabis and Cannabinoid Research*. 5 (2): 118-120. doi:10.1089/can.2020.0035.

Hindocha, C., T. P. Freeman, M. Grabski, J. B. Stroud, H. Crudgington, A. C. Davies, R. K. Das, W. Lawn, C. J. A. Morgan, and H. V. Curran. 2018. "Cannabidiol reverses attentional bias to cigarette cues in a human experimental model of tobacco withdrawal." *Addiction*. 113 (9): 1696-1705. doi:10.1111/add.14243.

Hindocha, C., T. P. Freeman, G. Schafer, C. Gardener, R. K. Das, C. J. Morgan, and H. V. Curran. 2015. "Acute effects of delta-9-tetrahydrocannabinol, cannabidiol and their combination on facial emotion recognition: A randomised, double-blind, placebo-controlled study in cannabis users." *Eur Neuropsychopharmacol* 25 (3): 325–334. doi:10.1016/j.euroneuro.2014.11.014.

Hine, B., E. Friedman, M. Torrelio, and S. Gershon. 1975. "Morphine-dependent rats: Blockade of precipitated abstinence by tetrahydrocannabinol." *Science* 187 (4175): 443–445. doi:10.1126/science.1167428.

Hine, B., M. Torrelio, and S. Gershon. 1975. "Interactions between cannabidiol and delta9-THC during abstinence in morphine-dependent rats." *Life Sci* 17 (6): 851–857. doi:10.1016/0024-3205(75)90435-x.

Hirvonen, J., R. S. Goodwin, C. T. Li, G. E. Terry, S. S. Zoghbi, C. Morse, V. W. Pike, N. D. Volkow, M. A. Huestis, and R. B. Innis. 2012. "Reversible and regionally selective downregulation of brain cannabinoid CB1 receptors in chronic daily cannabis smokers." *Mol Psychiatry* 17 (6): 642–649. doi:10.1038/mp.2011.82.

Hobbs, J. M., A. R. Vazquez, N. D. Remijan, R. E. Trotter, T. V. McMillan, K. E. Freedman, Y. Wei, K. A. Woelfel, O. R. Arnold, L. M. Wolfe, S. A. Johnson, and T. L. Weir. 2020. "Evaluation of pharmacokinetics and acute anti-inflammatory potential of two oral cannabidiol preparations in healthy adults." *Phytother Res*. 34 (7): 1696–1703. doi:10.1002/ptr.6651.

Hoggart, B., S. Ratcliffe, E. Ehler, K. H. Simpson, J. Hovorka, J. Lejcko, L. Taylor, H. Lauder, and M. Serpell. 2015. "A multicentre, open-label, follow-on study to assess the long-term maintenance of effect, tolerance and safety of THC/CBD oromucosal spray in the management of neuropathic pain." *J Neurol* 262 (1): 27–40. doi:10.1007/s00415-014-7502-9.

Holmes, A. 2001. "Targeted gene mutation approaches to the study of anxiety-like behavior in mice." *Neurosci Biobehav Rev* 25 (3): 261–273. doi:10.1016/s0149-7634(01)00012-4.

Howes, O. D., R. McCutcheon, M. J. Owen, and R. M. Murray. 2017. "The role of genes, stress, and dopamine in the development of schizophrenia." *Biol Psychiatry* 81 (1): 9–20. doi:10.1016/j.biopsych.2016.07.014.

Hsiao, Y. T., P. L. Yi, C. L. Li, and F. C. Chang. 2012. "Effect of cannabidiol on sleep disruption induced by the repeated combination tests consisting of open field and elevated plus-maze in rats." *Neuropharmacology* 62 (1): 373–384. doi:10.1016/j.neuropharm.2011.08.013.

Hudson, R., J. Renard, C. Norris, W. J. Rushlow, and S. R. Laviolette. 2019. "Cannabidiol counteracts the psychotropic side-effects of delta-9-tetrahydrocannabinol in the ventral hippocampus through bidirectional control of ERK1-2 phosphorylation." *J Neurosci* 39 (44): 8762–8777. doi:10.1523/JNEUROSCI.0708-19.2019.

Hundal, H., R. Lister, N. Evans, A. Antley, A. Englund, R. M. Murray, D. Freeman, and P. D. Morrison. 2018. "The effects of cannabidiol on persecutory ideation and anxiety in a high trait paranoid group." *J Psychopharmacol* 32 (3): 276–282. doi:10.1177/0269881117737400.

Hunt, C. A., R. T. Jones, R. I. Herning, and J. Bachman. 1981. "Evidence that cannabidiol does not significantly alter the pharmacokinetics of tetrahydrocannabinol in man." *J Pharmacokinet Biopharm* 9 (3): 245–260. doi:10.1007/bf01059266.

Hurd, Y. L., S. Spriggs, J. Alishayev, G. Winkel, K. Gurgov, C. Kudrich, A. M. Oprescu, and E. Salsitz. 2019. "Cannabidiol for the reduction of cue-induced craving and anxiety in drug-abstinent individuals with heroin use disorder: A double-blind randomized placebo-controlled trial." *Am J Psychiatry* 176 (11): 911–922. doi:10.1176/appi.ajp.2019.18101191.

Hurd, Y. L., M. Yoon, A. F. Manini, S. Hernandez, R. Olmedo, M. Ostman, and D. Jutras-Aswad. 2015. "Early phase in the development of cannabidiol as a treatment for addiction: Opioid relapse takes initial center stage." *Neurotherapeutics* 12 (4): 807–815. doi:10.1007/s13311-015-0373-7.

Iffland, K., and F. Grotenhermen. 2017. "An update on safety and side effects of cannabidiol: A review of clinical data and relevant animal studies." *Cannabis Cannabinoid Res* 2 (1): 139–154. doi:10.1089/can.2016.0034.

Ignatowska-Jankowska, B., M. M. Jankowski, and A. H. Swiergiel. 2011. "Cannabidiol decreases body weight gain in rats: Involvement of CB2 receptors." *Neurosci Lett* 490 (1): 82–84. doi:10.1016/j.neulet.2010.12.031.

Irving, P. M., T. Iqbal, C. Nwokolo, S. Subramanian, S. Bloom, N. Prasad, A. Hart, C. Murray, J. O. Lindsay, A. Taylor, R. Barron, and S. Wright. 2018. "A randomized, double-blind, placebo-controlled, parallel-group, pilot study of cannabidiol-rich botanical extract in the symptomatic treatment of ulcerative colitis." *Inflamm Bowel Dis* 24 (4): 714–724. doi:10.1093/ibd/izy002.

Izquierdo, I., O. A. Orsingher, and A. C. Berardi. 1973. "Effect of cannabidiol and of other *Cannabis sativa* compounds on hippocampal seizure discharges." *Psychopharmacologia* 28 (1): 95–102. doi:10.1007/bf00413961.

Izzo, A. A., F. Borrelli, R. Capasso, V. Di Marzo, and R. Mechoulam. 2009. "Non-psychotropic plant cannabinoids: New therapeutic opportunities from an ancient herb." *Trends Pharmacol Sci* 30 (10): 515–527. doi:10.1016/j.tips.2009.07.006.

Jacobs, A., and A. R. Todd. 1940. "Cannabidiol and cannabol, constituents of *Cannabis indica* resin." *Nature* 145: 350.

Jacobs, D. S., S. J. Kohut, S. Jiang, S. P. Nikas, A. Makriyannis, and J. Bergman. 2016. "Acute and chronic effects of cannabidiol on delta(9)-tetrahydrocannabinol (delta(9)-THC)-induced disruption in stop signal task performance." *Exp Clin Psychopharmacol* 24 (5): 320–330. doi:10.1037/pha0000081.

Jacobsen, L. K., S. M. Southwick, and T. R. Kosten. 2001. "Substance use disorders in patients with posttraumatic stress disorder: A review of the literature." *Am J Psychiatry* 158 (8): 1184–1190. doi:10.1176/appi.ajp.158.8.1184.

Jacobsson, S. O., E. Rongard, M. Stridh, G. Tiger, and C. J. Fowler. 2000. "Serum-dependent effects of tamoxifen and cannabinoids upon C6 glioma cell viability." *Biochem Pharmacol* 60 (12): 1807–1813. doi:10.1016/s0006-2952(00)00492-5.

Jamontt, J. M., A. Molleman, R. G. Pertwee, and M. E. Parsons. 2010. "The effects of delta-tetrahydrocannabinol and cannabidiol alone and in combination on damage, inflammation and in vitro motility disturbances in rat colitis." *Br J Pharmacol* 160 (3): 712–723. doi: 10.1111/j.1476-5381.2010.00791.x.

Jarocka-Karpowicz, I., M. Biernacki, A. Wronski, A. Gegotek, and E. Skrzydlewska. 2020. "Cannabidiol effects on phospholipid metabolism in keratinocytes from patients with psoriasis vulgaris." *Biomolecules* 10 (3): 367. doi:10.3390/biom10030367.

Jastrzab, A., A. Gegotek, and E. Skrzydlewska. 2019. "Cannabidiol regulates the expression of keratinocyte proteins involved in the inflammation process through transcriptional regulation." *Cells* 8 (8): 827. doi:10.3390/cells8080827.

Jefferson, D. A., H. E. Harding, S. O. Cawich, and A. Jackson-Gibson. 2013. "Postoperative analgesia in the Jamaican cannabis user." *J Psychoactive Drugs* 45 (3): 227–232. doi:10.1080/02791072.2013.803644.

Jennings, J. M., M. R. Angerame, C. L. Eschen, A. J. Phocas, and D. A. Dennis. 2019. "Cannabis use does not affect outcomes after total knee arthroplasty." *J Arthroplasty* 34 (8): 1667–1669. doi:10.1016/j.arth.2019.04.015.

Jesus, C. H. A., D. D. B. Redivo, A. T. Gasparin, B. B. Sotomaior, M. C. de Carvalho, K. Genaro, A. W. Zuardi, J. E. C. Hallak, J. A. Crippa, J. M. Zanoveli, and J. M. da Cunha. 2019. "Cannabidiol attenuates mechanical allodynia in streptozotocin-induced diabetic rats via serotonergic system activation through 5-HT1A receptors." *Brain Res* 1715: 156–164. doi:10.1016/j.brainres.2019.03.014.

Jhawar, N., E. Schoenberg, J. V. Wang, and N. Saedi. 2019. "The growing trend of cannabidiol in skincare products." *Clin Dermatol* 37 (3): 279–281. doi:10.1016/j.clindermatol.2018.11.002.

Jiang, R., S. Yamaori, Y. Okamoto, I. Yamamoto, and K. Watanabe. 2013. "Cannabidiol is a potent inhibitor of the catalytic activity of cytochrome P450 2C19." *Drug Metab Pharmacokinet* 28 (4): 332–338. doi:10.2133/dmpk.dmpk-12-rg-129.

Jiang, R., S. Yamaori, S. Takeda, I. Yamamoto, and K. Watanabe. 2011. "Identification of cytochrome P450 enzymes responsible for metabolism of cannabidiol by human liver microsomes." *Life Sci* 89 (5–6): 165–170. doi:10.1016/j.lfs.2011.05.018.

Johnson, J. R., M. Burnell-Nugent, D. Lossignol, E. D. Ganae-Motan, R. Potts, and M. T. Fallon. 2010. "Multicenter, double-blind, randomized, placebo-controlled, parallel-group study of the efficacy, safety, and tolerability of THC:CBD extract and THC extract in patients with intractable cancer-related pain." *J Pain Symptom Manage* 39 (2): 167–179. doi:10.1016/j.jpainsymman.2009.06.008.

Johnson, J. R., D. Lossignol, M. Burnell-Nugent, and M. T. Fallon. 2013. "An open-label extension study to investigate the long-term safety and tolerability of THC/CBD oromucosal spray and oromucosal THC spray in patients with terminal cancer-related pain refractory to strong opioid analgesics." *J Pain Symptom Manage* 46 (2): 207–218. doi:10.1016/j.jpainsymman.2012.07.014.

Jones, A. B., M. A. Elsohly, J. A. Bedford, and C. E. Turner. 1981. "Determination of cannabidiol in plasma by electron-capture gas chromatography." *J Chromatogr* 226 (1): 99–105. doi:10.1016/s0378-4347(00)84210-3.

Jones, G., and R. G. Pertwee. 1972. "A metabolic interaction in vivo between cannabidiol and 1-tetrahydrocannabinol." *Br J Pharmacol* 45 (2): 375–377. doi:10.1111/j.1476-5381.1972.tb08092.x.

Jones, N. A., A. J. Hill, I. Smith, S. A. Bevan, C. M. Williams, B. J. Whalley, and G. J. Stephens. 2010. "Cannabidiol displays antiepileptiform and antiseizure properties in vitro and in vivo." *J Pharmacol Exp Ther* 332 (2): 569–577. doi:10.1124/jpet.109.159145.

Jones, P. G., L. Flavello, O. Kennard, G. M. Sheldrick, and R. Mechoulam. 1977. "Cannabidiol." *Acta Cryst* B33: 3211–3214.

Juknat, A., E. Kozela, N. Kaushansky, R. Mechoulam, and Z. Vogel. 2016. "Anti-inflammatory effects of the cannabidiol derivative dimethylheptyl-cannabidiol—studies in BV-2 microglia and encephalitogenic T cells." *J Basic Clin Physiol Pharmacol* 27 (3): 289–296. doi:10.1515/jbcpp-2015-0071.

Jung, B., J. K. Lee, J. Kim, E. K. Kang, S. Y. Han, H. Y. Lee, and I. S. Choi. 2019. "Synthetic strategies for (−)-cannabidiol and its structural analogs." *Chem Asian J* 14 (21): 3749–3762. doi:10.1002/asia.201901179.

Jurkus, R., H. L. Day, F. S. Guimaraes, J. L. Lee, L. J. Bertoglio, and C. W. Stevenson. 2016. "Cannabidiol regulation of learned fear: Implications for treating anxiety-related disorders." *Front Pharmacol* 7: 454. doi:10.3389/fphar.2016.00454.

Kaczkurkin, A. N., and E. B. Foa. 2015. "Cognitive-behavioral therapy for anxiety disorders: An update on the empirical evidence." *Dialogues in Clinical Neuroscience* 17 (3): 337–346. doi:10.31887/DCNS.2015.17.3/akaczkurkin.

Kaplan, J. S., J. K. Wagner, K. Reid, F. McGuinness, S. Arvila, M. Brooks, H. Stevenson, J. Jones, B. Risch, T, McGillis, R. Budinich, E. Gambell, and B. Predovich. 2021. "Cannabidiol exposure during the mouse adolescent period is without harmful behavioral effects on locomotor activity, anxiety, and spatial memory." *Front Behav Neurosci* 15: 711639. doi:10.3389/fnbeh.2021.711639.

Kapur, S., and G. Remington. 2001. "Dopamine D(2) receptors and their role in atypical antipsychotic action: Still necessary and may even be sufficient." *Biol Psychiatry* 50 (11): 873–883. doi:10.1016/s0006-3223(01)01251-3.

Karler, R., and S. A. Turkanis. 1980. "Subacute cannabinoid treatment: anticonvulsant activity and withdrawal excitability in mice." *Br J Pharmacol* 68 (3): 479-484. doi:10.1111/j.1476-5381.1980.tb14562.x.

Karler, R., W. Cely, and S. A. Turkanis. 1973. "The anticonvulsant activity of cannabidiol and cannabinol." *Life Sci* 13 (11): 1527–1531. doi:10.1016/0024–3205(73)90141–0.

Karniol, I. G., and E. A. Carlini. 1973. "Pharmacological interaction between cannabidiol and δ9-tetrahydrocannabinol." *Psychopharmacologia* 33 (1): 53–70. doi:10.1007/BF00428793.

Karniol, I. G., I. Shirakawa, N. Kasinski, A. Pfeferman, and E. A. Carlini. 1974. "Cannabidiol interferes with the effects of delta 9—tetrahydrocannabinol in man." *Eur J Pharmacol* 28 (1): 172–177. doi:10.1016/0014-2999(74)90129-0.

Karschner, E. L., W. D. Darwin, R. S. Goodwin, S. Wright, and M. A. Huestis. 2011. "Plasma cannabinoid pharmacokinetics following controlled oral delta9-tetrahydrocannabinol and oromucosal cannabis extract administration." *Clin Chem* 57 (1): 66–75. doi:10.1373/clinchem.2010.152439.

Karschner, E. L., W. D. Darwin, R. P. McMahon, F. Liu, S. Wright, R. S. Goodwin, and M. A. Huestis. 2011. "Subjective and physiological effects after controlled Sativex and oral THC administration." *Clin Pharmacol Ther* 89 (3): 400–407. doi:10.1038/clpt.2010.318.

Katona, I. 2015. "Cannabis and endocannabinoid signaling in epilepsy." *Handb Exp Pharmacol* 231: 285–316. doi:10.1007/978-3-319-20825-1_10.

Katsidoni, V., I. Anagnostou, and G. Panagis. 2013. "Cannabidiol inhibits the reward-facilitating effect of morphine: Involvement of 5-HT1A receptors in the dorsal raphe nucleus." *Addict Biol* 18 (2): 286–296. doi:10.1111/j.1369-1600.2012.00483.x.

Katz-Talmor, D., S. Kivity, M. Blank, I. Katz, O. Perry, A. Volkov, I. Barshack, H. Amital, and Y. Shoenfeld. 2018. "Cannabidiol treatment in a murine model of systemic lupus erythematosus accelerates proteinuria development." *Isr Med Assoc J* 20 (12): 741–745.

Katzman, M. A., P. Bleau, P. Blier, P. Chokka, K. Kjernisted, M. Van Ameringen, M. M. Antony, S. Bouchard, A. Brunet, M. Flament, S. Grigoriadis, S. Mendlowitz, K. O'Connor, K. Rabheru, P. M. Richter, M. Robichaud, and J. R. Walker. 2014. "Canadian clinical practice guidelines for the management of anxiety, posttraumatic stress and obsessive-compulsive disorders." *BMC Psychiatry* 14 Suppl. 1: S1. doi:10.1186/1471-244x-14-s1-s1.

Kedzior, K. K., and L. T. Laeber. 2014. "A positive association between anxiety disorders and cannabis use or cannabis use disorders in the general population—a meta-analysis of 31 studies." *BMC Psychiatry* 14: 136. doi:10.1186/1471-244x-14-136.

Kenyon, J., W. Liu, and A. Dalgleish. 2018. "Report of objective clinical responses of cancer patients to pharmaceutical-grade synthetic cannabidiol." *Anticancer Res* 38 (10): 5831–5835. doi:10.21873/anticanres.12924.

Kessler, R. C., N. A. Sampson, P. Berglund, M. J. Gruber, A. Al-Hamzawi, L. Andrade, B. Bunting, K. Demyttenaere, S. Florescu, G. de Girolamo, O. Gureje, Y. He, C. Hu, Y. Huang, E. Karam, V. Kovess-Masfety, S. Lee, D. Levinson, M. E. Medina Mora, J. Moskalewicz, Y. Nakamura, F. Navarro-Mateu, M. A. Browne, M. Piazza, J. Posada-Villa, T. Slade, M. Ten Have, Y. Torres, G. Vilagut, M. Xavier, Z. Zarkov, V. Shahly, and M. A. Wilcox. 2015. "Anxious and non-anxious major depressive disorder in the World Health Organization World Mental Health Surveys." *Epidemiol Psychiatr Sci* 24 (3): 210–226. doi:10.1017/s2045796015000189.

Khodadadi, H., E. L. Salles, A. Jarrahi, F. Chibane, V. Costigliola, J. C. Yu, K. Vaibhav, D. C. Hess, K. M. Dhandapani, and B. Baban. 2020. "Cannabidiol modulates cytokine storm in acute respiratory distress syndrome induced by simulated viral infection using synthetic RNA." *Cannabis Cannabinoid Res* 5 (3): 197–201. doi:10.1089/can.2020.0043.

King, K. M., A. M. Myers, A. J. Soroka-Monzo, R. F. Tuma, R. J. Tallarida, E. A. Walker, and S. J. Ward. 2017. "Single and combined effects of delta(9)-tetrahydrocannabinol and cannabidiol in a mouse model of chemotherapy-induced neuropathic pain." *Br J Pharmacol* 174 (17): 2832–2841. doi:10.1111/bph.13887.

Kintz, P. 2021. "Vaping pure cannabidiol e-cigarettes does not produce detectable amount of 9-THC in human blood." *J Anal Toxicol.* 44 (9): e1–e2. doi:10.1093/jat/bkaa008.

Klein, C., E. Karanges, A. Spiro, A. Wong, J. Spencer, T. Huynh, N. Gunasekaran, T. Karl, L. E. Long, X. F. Huang, K. Liu, J. C. Arnold, and I. S. McGregor. 2011. "Cannabidiol potentiates delta(9)-tetrahydrocannabinol (THC) behavioural effects and alters THC pharmacokinetics during acute and chronic treatment in adolescent rats." *Psychopharmacology (Berl)* 218 (2): 443–457. doi:10.1007/s00213-011-2342-0.

Kogan, N. M., R. Rabinowitz, P. Levi, D. Gibson, P. Sandor, M. Schlesinger, and R. Mechoulam. 2004. "Synthesis and antitumor activity of quinonoid derivatives of cannabinoids." *J Med Chem* 47 (15): 3800–3806. doi:10.1021/jm040042o.

Kogan, N. M., M. Schlesinger, M. Peters, G. Marincheva, R. Beeri, and R. Mechoulam. 2007. "A cannabinoid anticancer quinone, HU-331, is more potent and less cardiotoxic than doxorubicin: A comparative in vivo study." *J Pharmacol Exp Ther* 322 (2): 646–653. doi:10.1124/jpet.107.120865.

Kogan, N. M., M. Schlesinger, E. Priel, R. Rabinowitz, E. Berenshtein, M. Chevion, and R. Mechoulam. 2007. "HU-331, a novel cannabinoid-based anticancer topoisomerase II inhibitor." *Mol Cancer Ther* 6 (1): 173–183. doi:10.1158/1535-7163.MCT-06-0039.

Koob, G. F., and N. Volkow. 2010. "Neurocircuitry of addiction." *Neuropsychopharmacology* 35 (1): 217–238. doi:10.1038/npp.2009.110.

Kosgodage, U. S., R. Mould, A. B. Henley, A. V. Nunn, G. W. Guy, E. L. Thomas, J. M. Inal, J. D. Bell, and S. Lange. 2018. "Cannabidiol (CBD) is a novel inhibitor for exosome and microvesicle (EMV) release in cancer." *Front Pharmacol* 9: 889. doi:10.3389/fphar.2018.00889.

Kosgodage, U. S., P. Uysal-Onganer, A. MacLatchy, R. Mould, A. V. Nunn, G. W. Guy, I. Kraev, N. P. Chatterton, E. L. Thomas, J. M. Inal, J. D. Bell, and S. Lange. 2019. "Cannabidiol affects extracellular vesicle release, miR21 and miR126, and reduces prohibitin protein in glioblastoma multiforme cells." *Transl Oncol* 12 (3): 513–522. doi:10.1016/j.tranon.2018.12.004.

Kosiba, J. D., S. A. Maisto, and J. W. Ditre. 2019. "Patient-reported use of medical cannabis for pain, anxiety, and depression symptoms: Systematic review and meta-analysis." *Soc Sci Med* 233: 181–192. doi:10.1016/j.socscimed.2019.06.005.

Kovalchuk, O., and I. Kovalchuk. 2020. "Cannabinoids as anticancer therapeutic agents." *Cell Cycle* 19 (9): 961–989. doi:10.1080/15384101.2020.1742952.

Kucerova, J., K. Tabiova, F. Drago, and V. Micale. 2014. "Therapeutic potential of cannabinoids in schizophrenia." *Recent Pat CNS Drug Discov* 9 (1): 13–25. doi:10.2174/1574889809666140307115532.

Kwiatkowska, M., L. A. Parker, P. Burton, and R. Mechoulam. 2004. "A comparative analysis of the potential of cannabinoids and ondansetron to suppress cisplatin-induced emesis in the *Suncus murinus* (house musk shrew)." *Psychopharmacology (Berl)* 174 (2): 254–259. doi:10.1007/s00213-003-1739-9.

Laczkovics, C., O. D. Kothgassner, A. Felnhofer, and C. M. Klier. 2020. "Cannabidiol treatment in an adolescent with multiple substance abuse, social anxiety and depression." *Neuropsychiatr* 35 (1): 31–34 doi:10.1007/s40211-020-00334-0.

Lafuente, H., F. J. Alvarez, M. R. Pazos, A. Alvarez, M. C. Rey-Santano, V. Mielgo, X. Murgia-Esteve, E. Hilario, and J. Martinez-Orgado. 2011. "Cannabidiol reduces brain damage and improves functional recovery after acute hypoxia-ischemia in newborn pigs." *Pediatr Res* 70 (3): 272–277. doi:10.1203/PDR.0b013e3182276b11.

Lafuente, H., M. R. Pazos, A. Alvarez, N. Mohammed, M. Santos, M. Arizti, F. J. Alvarez, and J. A. Martinez-Orgado. 2016. "Effects of cannabidiol and hypothermia on short-term brain damage in new-born piglets after acute hypoxia-ischemia." *Front Neurosci* 10: 323. doi:10.3389/fnins.2016.00323.

Lander, N., Z. Ben-Zvi, R. Mechoulam, B. Martin, M. Nordqvist, and S. Agurell. 1976. "Total synthesis of cannabidiol and delta1-tetrahydrocannabinol metabolites." *J Chem Soc Perkin* 1 (1): 8–16.

Langford, R. M., J. Mares, A. Novotna, M. Vachova, I. Novakova, W. Notcutt, and S. Ratcliffe. 2013. "A double-blind, randomized, placebo-controlled, parallel-group study of THC/CBD oromucosal spray in combination with the existing treatment regimen, in the relief of central neuropathic pain in patients with multiple sclerosis." *J Neurol* 260 (4): 984–997. doi:10.1007/s00415-012-6739-4.

Laprairie, R. B., A. M. Bagher, M. E. Kelly, and E. M. Denovan-Wright. 2015. "Cannabidiol is a negative allosteric modulator of the cannabinoid CB1 receptor." *Br J Pharmacol* 172 (20): 4790–4805. doi:10.1111/bph.13250.

Leas, Eric C., Erik M. Hendrickson, Alicia L. Nobles, Rory Todd, Davey M. Smith, Mark Dredze, and John W. Ayers. 2020. "Self-reported cannabidiol (CBD) use for conditions with proven therapies." *JAMA Network Open* 3 (10): e2020977–e2020977. doi:10.1001/jamanetworkopen.2020.20977.

Leas, E. C., A. L. Nobles, T. L. Caputi, M. Dredze, D. M. Smith, and J. W. Ayers. 2019. "Trends in Internet searches for cannabidiol (CBD) in the United States." *JAMA Netw Open* 2 (10): e1913853. doi:10.1001/jamanetworkopen.2019.13853.

LeDoux, J. E. 2000. "Emotion circuits in the brain." *Annu Rev Neurosci* 23: 155–184. doi:10.1146/annurev.neuro.23.1.155.

Lee, C. M., C. Neighbors, and B. A. Woods. 2007. "Marijuana motives: Young adults' reasons for using marijuana." *Addict Behav* 32 (7): 1384–1394. doi:10.1016/j.addbeh.2006.09.010.

Lee, C. Y., S. P. Wey, M. H. Liao, W. L. Hsu, H. Y. Wu, and T. R. Jan. 2008. "A comparative study on cannabidiol-induced apoptosis in murine thymocytes and EL-4 thymoma cells." *Int Immunopharmacol* 8 (5): 732–740. doi:10.1016/j.intimp.2008.01.018.

Lee, J. L. C., L. J. Bertoglio, F. S. Guimaraes, and C. W. Stevenson. 2017. "Cannabidiol regulation of emotion and emotional memory processing: Relevance for treating anxiety-related and substance abuse disorders." *Br J Pharmacol* 174 (19): 3242–3256. doi:10.1111/bph.13724.

Lee, W. S., K. Erdelyi, C. Matyas, P. Mukhopadhyay, Z. V. Varga, L. Liaudet, G. Hasku, D. Cihakova, R. Mechoulam, and P. Pacher. 2016. "Cannabidiol limits T cell-mediated chronic autoimmune myocarditis: Implications to autoimmune disorders and organ transplantation." *Mol Med* 22: 136–146. doi:10.2119/molmed.2016.00007.

Leehey, M. A., Y. Liu, F. Hart, C. Epstein, M. Cook, S. Sillau, J. Klawitter, H. Newman, C. Sempio, L. Forman, L. Seeberger, O. Klepitskaya, Z. Baud, and J. Bainbridge. 2020. "Safety and tolerability of cannabidiol in Parkinson disease: An open label, dose-escalation study." *Cannabis Cannabinoid Res* 5 (4): 326–336. doi:10.1089/can.2019.0068.

Lehmann, C., N. B. Fisher, B. Tugwell, A. Szczesniak, M. Kelly, and J. Zhou. 2016. "Experimental cannabidiol treatment reduces early pancreatic inflammation in type 1 diabetes." *Clin Hemorheol Microcirc* 64 (4): 655–662. doi:10.3233/CH-168021.

Leite, J. R., E. A. Carlini, N. Lander, and R. Mechoulam. 1982. "Anticonvulsant effects of the (–) and (+)isomers of cannabidiol and their dimethylheptyl homologs." *Pharmacology* 24 (3): 141–146. doi:10.1159/000137588.

Lemos, J. I., L. B. Resstel, and F. S. Guimaraes. 2010. "Involvement of the prelimbic prefrontal cortex on cannabidiol-induced attenuation of contextual conditioned fear in rats." *Behav Brain Res* 207 (1): 105–111. doi:10.1016/j.bbr.2009.09.045.

Levin, R., V. Almeida, F. F. Peres, M. B. Calzavara, N. D. da Silva, M. A. Suiama, S. T. Niigaki, A. W. Zuardi, J. E. Hallak, J. A. Crippa, and V. C. Abilio. 2012. "Antipsychotic profile of cannabidiol and rimonabant in an animal model of emotional context processing in schizophrenia." *Curr Pharm Des* 18 (32): 4960–4965. doi:10.2174/138161212802884735.

Levin, R., F. F. Peres, V. Almeida, M. B. Calzavara, A. W. Zuardi, J. E. Hallak, J. A. Crippa, and V. C. Abilio. 2014. "Effects of cannabinoid drugs on the deficit of prepulse inhibition of startle in an animal model of schizophrenia: The SHR strain." *Front Pharmacol* 5: 10. doi:10.3389/fphar.2014.00010.

Leweke, F. M., D. Piomelli, F. Pahlisch, D. Muhl, C. W. Gerth, C. Hoyer, J. Klosterkotter, M. Hellmich, and D. Koethe. 2012. "Cannabidiol enhances anandamide signaling and alleviates psychotic symptoms of schizophrenia." *Transl Psychiatry* 2: e94. doi:10.1038/tp.2012.15.

Leweke, F. M., C. Rohleder, C. W. Gerth, M. Hellmich, R. Pukrop, and D. Koethe. 2021. "Cannabidiol and amisulpride improve cognition in acute schizophrenia in an explorative, double-blind, active-controlled, randomized clinical trial." *Front Pharmacol* 12:614811. doi:10.3389/fphar.2021.614811.

Leweke, F. M., U. Schneider, M. Radwan, E. Schmidt, and H. M. Emrich. 2000. "Different effects of nabilone and cannabidiol on binocular depth inversion in man." *Pharmacol Biochem Behav* 66 (1): 175–181.

Li, H., W. Kong, C. R. Chambers, D. Yu, D. Ganea, R. F. Tuma, and S. J. Ward. 2018. "The non-psychoactive phytocannabinoid cannabidiol (CBD) attenuates pro-inflammatory mediators, T cell infiltration, and thermal sensitivity following spinal cord injury in mice." *Cell Immunol* 329: 1–9. doi:10.1016/j.cellimm.2018.02.016.

Lichtman, A. H., E. A. Lux, R. McQuade, S. Rossetti, R. Sanchez, W. Sun, S. Wright, E. Kornyeyeva, and M. T. Fallon. 2018. "Results of a double-blind, randomized, placebo-controlled study of nabiximols oromucosal spray as an adjunctive therapy in advanced cancer patients with chronic uncontrolled pain." *J Pain Symptom Manage* 55 (2): 179–188 e1. doi:10.1016/j.jpainsymman.2017.09.001.

Ligresti, A., L. De Petrocellis, and V. Di Marzo. 2016. "From phytocannabinoids to cannabinoid receptors and endocannabinoids: Pleiotropic physiological and pathological roles through complex pharmacology." *Physiol Rev* 96: 1593–1659. doi:10.1152/physrev.00002.2016.

Ligresti, A., A. S. Moriello, K. Starowicz, I. Matias, S. Pisanti, L. De Petrocellis, C. Laezza, G. Portella, M. Bifulco, and V. Di Marzo. 2006. "Antitumor activity of plant cannabinoids with emphasis on the effect of cannabidiol on human breast carcinoma." *J Pharmacol Exp Ther* 318 (3): 1375–1387. doi:10.1124/jpet.106.105247.

Lim, S. Y., S. Sharan, and S. Woo. 2020. "Model-based analysis of cannabidiol dose: Exposure relationship and bioavailability." *Pharmacotherapy* 40 (4): 291–300. doi:10.1002/phar.2377.

Limebeer, C. L., J. P. Krohn, S. Cross-Mellor, D. E. Litt, K. P. Ossenkopp, and L. A. Parker. 2008. "Exposure to a context previously associated with nausea elicits conditioned gaping in rats: A model of anticipatory nausea." *Behav Brain Res* 187 (1): 33–40. doi:10.1016/j.bbr.2007.08.024.

Limebeer, C. L., E. M. Rock, K. A. Sharkey, and L. A. Parker. 2018. "Nausea-induced 5-HT release in the interoceptive insular cortex and regulation by monoacylglycerol lipase (MAGL) inhibition and cannabidiol." *eNeuro* 5 (4): ENEURO.0256-18.2018. doi:10.1523/ENEURO.0256-18.2018.

Linares, I. M. P., F. S. Guimaraes, A. Eckeli, A. C. S. Crippa, A. W. Zuardi, J. D. S. Souza, J. E. Hallak, and J. A. S. Crippa. 2018. "No acute effects of cannabidiol on the sleep-wake cycle of healthy subjects: A randomized, double-blind, placebo-controlled, crossover study." *Front Pharmacol* 9: 315. doi:10.3389/fphar.2018.00315.

Linares, I. M., A. W. Zuardi, L. C. Pereira, R. H. Queiroz, R. Mechoulam, F. S. Guimaraes, and J. A. Crippa. 2019. "Cannabidiol presents an inverted

U-shaped dose-response curve in a simulated public speaking test." *Braz J Psychiatry* 41 (1): 9–14. doi:10.1590/1516-4446-2017-0015.

Linge, R., L. Jimenez-Sanchez, L. Campa, F. Pilar-Cuellar, R. Vidal, A. Pazos, A. Adell, and A. Diaz. 2016. "Cannabidiol induces rapid-acting antidepressant-like effects and enhances cortical 5-HT/glutamate neurotransmission: role of 5-HT1A receptors." *Neuropharmacology* 103: 16–26. doi:10.1016/j.neuropharm .2015.12.017.

Linher-Melville, K., Y. F. Zhu, J. Sidhu, N. Parzei, A. Shahid, G. Seesankar, D. Ma, Z. Wang, N. Zacal, M. Sharma, V. Parihar, R. Zacharias, and G. Singh. 2020. "Evaluation of the preclinical analgesic efficacy of naturally derived, orally administered oil forms of Δ9-tetrahydrocannabinol (THC), cannabidiol (CBD), and their 1:1 combination." *PLoS One* 15 (6): e0234176. doi:10.1371 /journal.pone.0234176.

Lintzeris, N., A. Bhardwaj, L. Mills, A. Dunlop, J. Copeland, I. McGregor, R. Bruno, J. Gugusheff, N. Phung, M. Montebello, T. Chan, A. Kirby, M. Hall, M. Jefferies, J. Luksza, M. Shanahan, R. Kevin, D. Allsop, and Group Agonist Replacement for Cannabis Dependence study. 2019. "Nabiximols for the treatment of cannabis dependence: A randomized clinical trial." *JAMA Intern Med.* 179 (9): 1242–1253. doi:10.1001/jamainternmed.2019.1993.

Liput, D. J., D. C. Hammell, A. L. Stinchcomb, and K. Nixon. 2013. "Transdermal delivery of cannabidiol attenuates binge alcohol-induced neurodegeneration in a rodent model of an alcohol use disorder." *Pharmacol Biochem Behav* 111: 120–127. doi:10.1016/j.pbb.2013.08.013.

Lodzki, M., B. Godin, L. Rakou, R. Mechoulam, R. Gallily, and E. Touitou. 2003. "Cannabidiol-transdermal delivery and anti-inflammatory effect in a murine model." *J Control Release* 93 (3): 377–387. doi:10.1016/ j.jconrel.2003.09.001.

Long, L. E., R. Chesworth, X. F. Huang, I. S. McGregor, J. C. Arnold, and T. Karl. 2010. "A behavioural comparison of acute and chronic delta9-tetrahydrocannabinol and cannabidiol in C57BL/6JArc mice." *Int J Neuropsychopharmacol* 13 (7): 861–876. doi:10.1017/S1461145709990605.

Long, L. E., D. T. Malone, and D. A. Taylor. 2006. "Cannabidiol reverses MK-801-induced disruption of prepulse inhibition in mice." *Neuropsychopharmacology* 31 (4): 795–803. doi:10.1038/sj.npp.1300838.

Lopez-Sendon Moreno, J. L., J. Garcia Caldentey, P. Trigo Cubillo, C. Ruiz Romero, G. Garcia Ribas, M. A. Alonso Arias, M. J. Garcia de Yebenes,

R. M. Tolon, I. Galve-Roperh, O. Sagredo, S. Valdeolivas, E. Resel, S. Ortega-Gutierrez, M. L. Garcia-Bermejo, J. Fernandez Ruiz, M. Guzman, and J. Garcia de Yebenes Prous. 2016. "A double-blind, randomized, crossover, placebo-controlled, pilot trial with Sativex in Huntington's disease." *J Neurol* 263 (7): 1390–1400. doi:10.1007/s00415-016-8145-9.

Lowe, H. I., N. J. Toyang, and W. McLaughlin. 2017. "Potential of cannabidiol for the treatment of viral hepatitis." *Pharmacognosy Res* 9 (1): 116–118. doi:10.4103/0974-8490.199780.

Lujan, M. A., A. Castro-Zavala, L. Alegre-Zurano, and O. Valverde. 2018. "Repeated Cannabidiol treatment reduces cocaine intake and modulates neural proliferation and CB1R expression in the mouse hippocampus." *Neuropharmacology* 143: 163–175. doi:10.1016/j.neuropharm.2018.09.043.

Machado Rocha, F. C., S. C. Stefano, R. De Cassia Haiek, L. M. Rosa Oliveira, and D. X. Da Silveira. 2008. "Therapeutic use of *Cannabis sativa* on chemotherapy-induced nausea and vomiting among cancer patients: Systematic review and meta-analysis." *European Journal of Cancer Care* 17 (5): 431–443. doi:10.1111/j.1365-2354.2008.00917.x.

Madden, K., K. Tanco, and E. Bruera. 2020. "Clinically significant drug-drug interaction between methadone and cannabidiol." *Pediatrics* 145 (6): e20193256. doi:10.1542/peds.2019-3256.

Magen, I., Y. Avraham, Z. Ackerman, L. Vorobiev, R. Mechoulam, and E. M. Berry. 2009. "Cannabidiol ameliorates cognitive and motor impairments in mice with bile duct ligation." *J Hepatol* 51 (3): 528–534. doi:10.1016/j.jhep.2009.04.021.

Mahmud, A., S. Gallant, F. Sedki, T. D'Cunha, and U. Shalev. 2017. "Effects of an acute cannabidiol treatment on cocaine self-administration and cue-induced cocaine seeking in male rats." *J Psychopharmacol* 31 (1): 96–104. doi:10.1177/0269881116667706.

Maida, V., and P. J. Daeninck. 2016. "A user's guide to cannabinoid therapies in oncology." *Curr Oncol* 23 (6): 398–406. doi:10.3747/co.23.3487.

Maione, S., F. Piscitelli, L. Gatta, D. Vita, L. De Petrocellis, E. Palazzo, V. de Novellis, and V. Di Marzo. 2011. "Non-psychoactive cannabinoids modulate the descending pathway of antinociception in anaesthetized rats through several mechanisms of action." *Br J Pharmacol* 162 (3): 584–596. doi:10.1111/j.1476-5381.2010.01063.x.

Malfait, A. M., R. Gallily, P. F. Sumariwalla, A. S. Malik, E. Andreakos, R. Mechoulam, and M. Feldmann. 2000. "The nonpsychoactive cannabis constituent cannabidiol is an oral anti-arthritic therapeutic in murine collagen-induced arthritis." *Proc Natl Acad Sci USA* 97 (17): 9561–9566. doi:10.1073/pnas.160105897.

Malinowska, B., M. Baranowska-Kuczko, A. Kicman, and E. Schlicker. 2021. "Opportunities, challenges and pitfalls of using cannabidiol as an adjuvant drug in COVID-19." *Int J Mol Sci* 22 (4): 1986. doi:10.3390/ijms22041986.

Malone, D. T., D. Jongejan, and D. A. Taylor. 2009. "Cannabidiol reverses the reduction in social interaction produced by low dose delta(9)-tetrahydrocannabinol in rats." *Pharmacol Biochem Behav* 93 (2): 91–96. doi:10.1016/j.pbb.2009.04.010.

Mamber, S. W., S. Krakowka, J. Osborn, L. Saberski, R. G. Rhodes, A. E. Dahlberg, S. Pond-Tor, K. Fitzgerald, N. Wright, S. Beseme, and J. McMichael. 2020. "Can unconventional immunomodulatory agents help alleviate COVID-19 symptoms and severity?" *mSphere* 5 (3): e00288-20. doi:10.1128/mSphere.00288-20.

Manini, A. F., G. Yiannoulos, M. M. Bergamaschi, S. Hernandez, R. Olmedo, A. J. Barnes, G. Winkel, R. Sinha, D. Jutras-Aswad, M. A. Huestis, and Y. L. Hurd. 2015. "Safety and pharmacokinetics of oral cannabidiol when administered concomitantly with intravenous fentanyl in humans." *J Addict Med* 9 (3): 204–210. doi:10.1097/ADM.0000000000000118.

Mao, J., D. D. Price, J. Lu, L. Keniston, and D. J. Mayer. 2000. "Two distinctive antinociceptive systems in rats with pathological pain." *Neurosci Lett* 280 (1): 13–16. doi:10.1016/s0304-3940(99)00998-2.

Maren, S., and G. J. Quirk. 2004. "Neuronal signalling of fear memory." *Nat Rev Neurosci* 5 (11): 844–852. doi:10.1038/nrn1535.

Markova, J., U. Essner, B. Akmaz, M. Marinelli, C. Trompke, A. Lentschat, and C. Vila. 2019. "Sativex as add-on therapy versus further optimized first-line antispastics (SAVANT) in resistant multiple sclerosis spasticity: A double-blind, placebo-controlled randomised clinical trial." *Int J Neurosci* 129 (2): 119–128. doi:10.1080/00207454.2018.1481066.

Martell, K., A. Fairchild, B. LeGerrier, R. Sinha, S. Baker, H. Liu, A. Ghose, I. A. Olivotto, and M. Kerba. 2018. "Rates of cannabis use in patients with cancer." *Current Oncology* 25 (3): 219-225. doi:10.3747/co.25.3983.

Martin, B. R., D. J. Harvey, and W. D. Paton. 1977. "Biotransformation of cannabidiol in mice: Identification of new acid metabolites." *Drug Metab Dispos* 5 (3): 259–267.

Martin-Moreno, A. M., D. Reigada, B. G. Ramirez, R. Mechoulam, N. Innamorato, A. Cuadrado, and M. L. de Ceballos. 2011. "Cannabidiol and other cannabinoids reduce microglial activation in vitro and in vivo: Relevance to Alzheimer's disease." *Mol Pharmacol* 79 (6): 964–973. doi:10.1124/mol.111.071290.

Martin-Santos, R., J. A. Crippa, A. Batalla, S. Bhattacharyya, Z. Atakan, S. Borgwardt, P. Allen, M. Seal, K. Langohr, M. Farre, A. W. Zuardi, and P. K. McGuire. 2012. "Acute effects of a single, oral dose of d9-tetrahydrocannabinol (THC) and cannabidiol (CBD) administration in healthy volunteers." *Curr Pharm Des* 18 (32): 4966–4979.

Martinez-Pinilla, E., K. Varani, I. Reyes-Resina, E. Angelats, F. Vincenzi, C. Ferreiro-Vera, J. Oyarzabal, E. I. Canela, J. L. Lanciego, X. Nadal, G. Navarro, P. A. Borea, and R. Franco. 2017. "Binding and signaling studies disclose a potential allosteric site for cannabidiol in cannabinoid CB2 receptors." *Front Pharmacol* 8: 744. doi:10.3389/fphar.2017.00744.

Masataka, N. 2019. "Anxiolytic effects of repeated cannabidiol treatment in teenagers with social anxiety disorders." *Front Psychol* 10: 2466. doi:10.3389/fpsyg.2019.02466.

Massi, P., M. Solinas, V. Cinquina, and D. Parolaro. 2013. "Cannabidiol as potential anticancer drug." *Br J Clin Pharmacol* 75 (2): 303–312. doi:10.1111/j.1365-2125.2012.04298.x.

Massi, P., A. Vaccani, S. Bianchessi, B. Costa, P. Macchi, and D. Parolaro. 2006. "The non-psychoactive cannabidiol triggers caspase activation and oxidative stress in human glioma cells." *Cell Mol Life Sci* 63 (17): 2057–2066. doi:10.1007/s00018-006-6156-x.

Massi, P., A. Vaccani, S. Ceruti, A. Colombo, M. P. Abbracchio, and D. Parolaro. 2004. "Antitumor effects of cannabidiol, a nonpsychoactive cannabinoid, on human glioma cell lines." *J Pharmacol Exp Ther* 308 (3): 838–845. doi:10.1124/jpet.103.061002.

Mbachi, C., B. Attar, O. Oyenubi, W. Yuchen, A. Efesomwan, I. Paintsil, M. Madhu, O. Ajiboye, C. R. Simons-Linares, W. E. Trick, and V. Kotwal. 2019. "Association between cannabis use and complications related to ulcerative colitis in hospitalized patients: A propensity matched retrospective cohort study." *Medicine (Baltimore)* 98 (32): e16551. doi:10.1097/md.0000000000016551.

McAllister, S. D., R. T. Christian, M. P. Horowitz, A. Garcia, and P. Y. Desprez. 2007. "Cannabidiol as a novel inhibitor of Id-1 gene expression in aggressive breast cancer cells." *Mol Cancer Ther* 6 (11): 2921–2927. doi:10.1158/1535-7163.MCT-07-0371.

McCartney, D., R. C. Kevin, A.S.Suraev, C. Irwin, R.R. Grunstein, C. M. Hoyos, I. McGregor. 2021. "Orally administered cannabidiol does not produce false-positive tests for $\Delta^9$-tetrahydrocannabinol on the Securetec DrugWipe® 5S or Dräger DrugTest® 5000." *Drug Testing and Analysis*, 1–7. doi:10.1002/dta.3153.

McGuire, P., P. Robson, W. J. Cubala, D. Vasile, P. D. Morrison, R. Barron, A. Taylor, and S. Wright. 2018. "Cannabidiol (CBD) as an adjunctive therapy in schizophrenia: A multicenter randomized controlled trial." *Am J Psychiatry* 175 (3): 225–231. doi:10.1176/appi.ajp.2017.17030325.

McKallip, R. J., W. Jia, J. Schlomer, J. W. Warren, P. S. Nagarkatti, and M. Nagarkatti. 2006. "Cannabidiol-induced apoptosis in human leukemia cells: A novel role of cannabidiol in the regulation of p22phox and Nox4 expression." *Mol Pharmacol* 70 (3): 897–908. doi:10.1124/mol.106.023937.

Mechoulam, R. 1986. "Interview with Prof. Raphael Mechoulam, codiscoverer of THC: Interview by Stanley Einstein." *Int J Addict* 21 (4–5): 579–587. doi:10.3109/10826088609083542.

Mechoulam, R., and E. A. Carlini. 1978. "Toward drugs derived from cannabis." *Naturwissenschaften* 65 (4): 174–179. doi:10.1007/bf00450585.

Mechoulam, R., and Y. Gaoni. 1967a. "The absolute configuration of delta-1-tetrahydrocannabinol, the major active constituent of hashish." *Tetrahedron Lett* 12: 1109–1111. doi:10.1016/s0040-4039(00)90646-4.

Mechoulam, R., and Y. Gaoni. 1967b. "Recent advances in the chemistry of hashish." *Fortschr Chem Org Naturst* 25: 175–213. doi:10.1007/978-3-7091-8164-5_6.

Mechoulam, R., and L. Hanus. 2002. "Cannabidiol: An overview of some chemical and pharmacological aspects. Part I: Chemical aspects." *Chem Phys Lipids* 121 (1–2): 35–43. doi:10.1016/s0009-3084(02)00144-5.

Mechoulam, R., L. O. Hanus, R. Pertwee, and A. C. Howlett. 2014. "Early phytocannabinoid chemistry to endocannabinoids and beyond." *Nat Rev Neurosci* 15 (11): 757–764. doi:10.1038/nrn3811.

Mechoulam, R., and L. A. Parker. 2013. "The endocannabinoid system and the brain." *Annu Rev Psychol* 64: 21–47. doi:10.1146/annurev-psych-113011-143739.

Mechoulam, R., L. A. Parker, and R. Gallily. 2002. "Cannabidiol: An overview of some pharmacological aspects." *J Clin Pharmacol* 42 (S1): 11S-19S. doi:10.1002/j.1552-4604.2002.tb05998.x.

Mechoulam, R., and Y. Shvo. 1963. "Hashish. I. The structure of cannabidiol." *Tetrahedron* 19 (12): 2073–2078. doi:10.1016/0040-4020(63)85022-x.

Mehta, P., D. F. McAuley, M. Brown, E. Sanchez, R. S. Tattersall, and J. J. Manson. 2020. "COVID-19: Consider cytokine storm syndromes and immunosuppression." *Lancet* 395 (10229): 1033–1034. doi:10.1016/S0140-6736(20)30628-0.

Mejia, S., F. M. Duerr, G. Griffenhagen, and S. McGrath. 2021. "Evaluation of the effect of cannabidiol on naturally occurring osteoarthritis-associated pain: A pilot study in dogs." *J Am Anim Hosp Assoc* 57 (2): 81–90. doi:10.5326/jaaha-ms-7119.

Melas, P. A., M. Scherma, W. Fratta, C. Cifani, and P. Fadda. 2021. "Cannabidiol as a potential treatment for anxiety and mood disorders: Molecular targets and epigenetic insights from preclinical research." *Int J Mol Sci* 22 (4): 1863. doi:10.3390/ijms22041863.

Merrick, J., B. Lane, T. Sebree, T. Yaksh, C. O'Neill, and S. L. Banks. 2016. "Identification of psychoactive degradants of cannabidiol in simulated gastric and physiological fluid." *Cannabis Cannabinoid Res* 1 (1): 102–112. doi:10.1089/can.2015.0004.

Meuth, S. G., T. Henze, U. Essner, C. Trompke, and C. Vila Silvan. 2020. "Tetrahydrocannabinol and cannabidiol oromucosal spray in resistant multiple sclerosis spasticity: Consistency of response across subgroups from the SAVANT randomized clinical trial." *Int J Neurosci* 130 (12): 1199–1205. doi:10.1080/00207454.2020.1730832.

Miller, I., I. E. Scheffer, B. Gunning, R. Sanchez-Carpintero, A. Gil-Nagel, M. S. Perry, R. P. Saneto, D. Checketts, E. Dunayevich, V. Knappertz, and

Gwpcare Study Group. 2020. "Dose-ranging effect of adjunctive oral cannabidiol vs placebo on convulsive seizure frequency in Dravet syndrome: A randomized clinical trial." *JAMA Neurol.* 77( 5): 613–621. doi:10.1001/jamaneurol.2020.0073.

Mishima, K., K. Hayakawa, K. Abe, T. Ikeda, N. Egashira, K. Iwasaki, and M. Fujiwara. 2005. "Cannabidiol prevents cerebral infarction via a serotonergic 5-hydroxytryptamine1A receptor-dependent mechanism." *Stroke* 36 (5): 1077–1082. doi:10.1161/01.STR.0000163083.59201.34.

Mitchell, V. A., J. Harley, S. L. Casey, A. C. Vaughan, B. L. Winters, and C. W. Vaughan. 2021. "Oral efficacy of Δ(9)-tetrahydrocannabinol and cannabidiol in a mouse neuropathic pain model." *Neuropharmacology* 189: 108529. doi:10.1016/j.neuropharm.2021.108529.

Mlost, J., M. Bryk, and K. Starowicz. 2020. "Cannabidiol for pain treatment: Focus on pharmacology and mechanism of action." *Int J Mol Sci* 21 (22): 8870. doi:10.3390/ijms21228870.

Mohammed, N., M. Ceprian, L. Jimenez, M. R. Pazos, and J. Martinez-Orgado. 2017. "Neuroprotective effects of cannabidiol in hypoxic ischemic insult: The therapeutic window in newborn mice." *CNS Neurol Disord Drug Targets* 16 (1): 102–108. doi:10.2174/1871527315666160927110305.

Mondello, E., D. Quattrone, L. Cardia, G. Bova, R. Mallamace, A. A. Barbagallo, C. Mondello, C. Mannucci, M. Di Pietro, V. Arcoraci, and G. Calapai. 2018. "Cannabinoids and spinal cord stimulation for the treatment of failed back surgery syndrome refractory pain." *J Pain Res* 11: 1761–1767. doi:10.2147/JPR.S166617.

Mongeau-Pérusse, V., S. Brissette, J. Bruneau, P. Conrod, S. Dubreucq, G. Gazil, E. Stip, and D. Jutras-Aswad. 2021. "Cannabidiol as a treatment for craving and relapse in individuals with cocaine use disorder: A randomized placebo-controlled trial." *Addiction.* 116 (9): 2431–2442. doi:10.1111/add.15417.

Montgomery, K. C. 1955. "The relation between fear induced by novel stimulation and exploratory behavior." *J Comp Physiol Psychol* 48 (4): 254–260. doi:10.1037/h0043788.

Monti, J. M. 1977. "Hypnoticlike effects of cannabidiol in the rat." *Psychopharmacology (Berl)* 55 (3): 263–265. doi:10.1007/bf00497858.

Montoya, Z. T., A. L. Uhernik, and J. P. Smith. 2020. "Comparison of cannabidiol to citalopram in targeting fear memory in female mice." *J Cannabis Res* 2 (1): 48. doi:10.1186/s42238-020-00055-9.

Moreira, F. A., and F. S. Guimaraes. 2005. "Cannabidiol inhibits the hyperlocomotion induced by psychotomimetic drugs in mice." *Eur J Pharmacol* 512 (2–3): 199–205. doi:10.1016/j.ejphar.2005.02.040.

Morgan, C. J., and H. V. Curran. 2008. "Effects of cannabidiol on schizophrenia-like symptoms in people who use cannabis." *Br J Psychiatry* 192 (4): 306–307. doi:10.1192/bjp.bp.107.046649.

Morgan, C. J., R. K. Das, A. Joye, H. V. Curran, and S. K. Kamboj. 2013. "Cannabidiol reduces cigarette consumption in tobacco smokers: Preliminary findings." *Addict Behav* 38 (9): 2433–2436. doi:10.1016/j.addbeh.2013.03.011.

Morgan, C. J. A., T. P. Freeman, C. Hindocha, G. Schafer, C. Gardner, and H. V. Curran. 2018. "Individual and combined effects of acute delta-9-tetrahydrocannabinol and cannabidiol on psychotomimetic symptoms and memory function." *Transl Psychiatry* 8 (1): 181. doi:10.1038/s41398-018-0191-x.

Morgan, C. J., T. P. Freeman, G. L. Schafer, and H. V. Curran. 2010. "Cannabidiol attenuates the appetitive effects of delta 9-tetrahydrocannabinol in humans smoking their chosen cannabis." *Neuropsychopharmacology* 35 (9): 1879–1885. doi:10.1038/npp.2010.58.

Morgan, C. J., C. Gardener, G. Schafer, S. Swan, C. Demarchi, T. P. Freeman, P. Warrington, I. Rupasinghe, A. Ramoutar, N. Tan, G. Wingham, S. Lewis, and H. V. Curran. 2012. "Sub-chronic impact of cannabinoids in street cannabis on cognition, psychotic-like symptoms and psychological well-being." *Psychol Med* 42 (2): 391–400. doi:10.1017/S0033291711001322.

Morgan, C. J., G. Schafer, T. P. Freeman, and H. V. Curran. 2010. "Impact of cannabidiol on the acute memory and psychotomimetic effects of smoked cannabis: Naturalistic study: naturalistic study [corrected]." *Br J Psychiatry* 197 (4): 285–290. doi:10.1192/bjp.bp.110.077503.

Morissette, F., V. Mongeau-Perusse, E. Rizkallah, R. Thebault, S. Lepage, S. Brissette, J. Bruneau, S. Dubreucq, E. Stip, J.F. Cailhier, D. Jutras-Aswad. 2021. "Exploring cannabidiol effects on inflammatory marker in individuals with cocaine use disorder: a randomized controlled trial." *Neuropsychopharmacology* 46 (12): 2101–2111. doi:10.1038/s41386-021-01098-z

Moulin, D. E., A. J. Clark, I. Gilron, M. A. Ware, C. P. Watson, B. J. Sessle, T. Coderre, P. K. Morley-Forster, J. Stinson, A. Boulanger, P. Peng, G. A. Finley, P. Taenzer, P. Squire, D. Dion, A. Cholkan, A. Gilani, A. Gordon, J. Henry, R. Jovey, M. Lynch, A. Mailis-Gagnon, A. Panju, G. B. Rollman, and A. Velly. 2007. "Pharmacological management of chronic neuropathic pain—consensus statement and guidelines from the Canadian Pain Society." *Pain Res Manag* 12 (1): 13–21. doi:10.1155/2007/730785.

Mukhopadhyay, P., M. Rajesh, B. Horvath, S. Batkai, O. Park, G. Tanchian, R. Y. Gao, V. Patel, D. A. Wink, L. Liaudet, G. Hasko, R. Mechoulam, and P. Pacher. 2011. "Cannabidiol protects against hepatic ischemia/reperfusion injury by attenuating inflammatory signaling and response, oxidative/nitrative stress, and cell death." *Free Radic Biol Med* 50 (10): 1368–1381. doi:10.1016/j.freeradbiomed.2011.02.021.

Murillo-Rodriguez, E., G. Arankowsky-Sandoval, N. B. Rocha, R. Peniche-Amante, A. B. Veras, S. Machado, and H. Budde. 2018. "Systemic injections of cannabidiol enhance acetylcholine levels from basal forebrain in rats." *Neurochem Res* 43 (8): 1511–1518. doi:10.1007/s11064-018-2565-0.

Murillo-Rodriguez, E., G. Arankowsky-Sandoval, R. G. Pertwee, L. Parker, and R. Mechoulam. 2020. "Sleep and neurochemical modulation by cannabidiolic acid methyl ester in rats." *Brain Res Bull* 155: 166–173. doi:10.1016/j.brainresbull.2019.12.006.

Murillo-Rodríguez, E., D. Millán-Aldaco, D. Cicconcelli, V. Giorgetti, G. Arankowsky-Sandoval, J. Alcaraz-Silva, C. Imperatori, S. Machado, H. Budde, and P. Torterolo. 2021. "Sleep-wake cycle disturbances and NeuN-altered expression in adult rats after cannabidiol treatments during adolescence." *Psychopharmacology (Berl)*. 238 (6): 1437–1447. doi:10.1007/s00213-021-05769-z.

Murillo-Rodriguez, E., D. Millan-Aldaco, M. Palomero-Rivero, R. Mechoulam, and R. Drucker-Colin. 2006. "Cannabidiol, a constituent of *Cannabis sativa*, modulates sleep in rats." *FEBS Lett* 580 (18): 4337–4435. doi:10.1016/j.febslet.2006.04.102.

Murillo-Rodriguez, E., D. Millan-Aldaco, M. Palomero-Rivero, R. Mechoulam, and R. Drucker-Colin. 2008. "The nonpsychoactive cannabis constituent cannabidiol is a wake-inducing agent." *Behav Neurosci* 122 (6): 1378–1382. doi:10.1037/a0013278.

Murillo-Rodriguez, E., D. Millan-Aldaco, M. Palomero-Rivero, D. Morales-Lara, R. Mechoulam, and R. Drucker-Colin. 2019. "Cannabidiol partially blocks the excessive sleepiness in hypocretindeficient rats: Preliminary data." *CNS Neurol Disord Drug Targets* 18 (9): 705–712. doi:10.2174/1871527 318666191021143300.

Murillo-Rodriguez, E., M. Palomero-Rivero, D. Millan-Aldaco, R. Mechoulam, and R. Drucker-Colin. 2011. "Effects on sleep and dopamine levels of microdialysis perfusion of cannabidiol into the lateral hypothalamus of rats." *Life Sci* 88 (11–12): 504–511. doi:10.1016/j.lfs.2011.01.013.

Murillo-Rodriguez, E., A. Sarro-Ramirez, D. Sanchez, S. Mijangos-Moreno, A. Tejeda-Padron, A. Poot-Ake, K. Guzman, E. Pacheco-Pantoja, and O. Arias-Carrion. 2014. "Potential effects of cannabidiol as a wake-promoting agent." *Curr Neuropharmacol* 12 (3): 269–272. doi:10.2174/1570159X11666131204235805.

Murkar, A., P. Kent, C. Cayer, J. James, T. Durst, and Z. Merali. 2019. "Cannabidiol and the remainder of the plant extract modulate the effects of delta9-tetrahydrocannabinol on fear memory reconsolidation." *Front Behav Neurosci* 13: 174. doi:10.3389/fnbeh.2019.00174.

Murphy, M., S. Mills, J. Winstone, E. Leishman, J. Wager-Miller, H. Bradshaw, and K. Mackie. 2017. "Chronic adolescent delta(9)-tetrahydrocannabinol treatment of male mice leads to long-term cognitive and behavioral dysfunction, which are prevented by concurrent cannabidiol treatment." *Cannabis Cannabinoid Res* 2 (1): 235–246. doi:10.1089/can.2017.0034.

Nadulski, T., F. Pragst, G. Weinberg, P. Roser, M. Schnelle, E. M. Fronk, and A. M. Stadelmann. 2005. "Randomized, double-blind, placebo-controlled study about the effects of cannabidiol (CBD) on the pharmacokinetics of delta9-tetrahydrocannabinol (THC) after oral application of THC verses standardized cannabis extract." *Ther Drug Monit* 27 (6): 799–810. doi:10.1097/01.ftd.0000177223.19294.5c.

Naftali, T. 2020. "An overview of cannabis-based treatment in Crohn's disease." *Expert Rev Gastroenterol Hepatol* 14 (4): 253–257. doi:10.1080/17474124.2020.1740590.

Naftali, T., L. Bar-Lev Schleider, I. Dotan, E. P. Lansky, F. Skleroversusky Benjaminov, and F. M. Konikoff. 2013. "Cannabis induces a clinical response in patients with Crohn's disease: A prospective placebo-controlled study." *Clin Gastroenterol Hepatol* 11 (10): 1276–1280 e1. doi:10.1016/j.cgh.2013.04.034.

Naftali, T., L. Bar-Lev Schleider, S. Almog, D. Meiri, and F. M. Konikoff. 2021. "Oral CBD-rich cannabis induces clinical but not endoscopic response in patients with Crohn's disease, a randomised controlled trial." *Journal of Crohn's and Colitis* 15 (11): 1799–1806. doi:10.1093/ecco-jcc/jjab069.

Naftali, T., R. Mechulam, A. Marii, G. Gabay, A. Stein, M. Bronshtain, I. Laish, F. Benjaminov, and F. M. Konikoff. 2017. "Low-dose cannabidiol is safe but not effective in the treatment for Crohn's disease, a randomized controlled trial." *Dig Dis Sci* 62 (6): 1615–1620. doi:10.1007/s10620-017-4540-z.

Nahler, G., F. Grotenhermen, A. W. Zuardi, and J. A. S. Crippa. 2017. "A conversion of oral cannabidiol to delta9-tetrahydrocannabinol seems not to occur in humans." *Cannabis Cannabinoid Res* 2 (1): 81–86. doi:10.1089/can.2017.0009.

National Academies of Sciences, Engineering, and Medicine; Health and Medicine Division; Board on Population Health and Public Health Practice; Committee on the Health Effects of Marijuana. 2017. *The Health Effects of Cannabis and Cannabinoids: The Current State of Evidence and Recommendations for Research*. Washington, DC: National Academies Press.

Navari, R. M., and M. Aapro. 2016. "Antiemetic prophylaxis for chemotherapy-induced nausea and vomiting." *N Engl J Med* 374 (14): 1356–1367. doi:10.1056/NEJMra1515442.

Neelakantan, H., R. J. Tallarida, Z. W. Reichenbach, R. F. Tuma, S. J. Ward, and E. A. Walker. 2015. "Distinct interactions of cannabidiol and morphine in three nociceptive behavioral models in mice." *Behav Pharmacol* 26 (3): 304–314. doi:10.1097/FBP.0000000000000119.

Nestler, E. J., and S. E. Hyman. 2010. "Animal models of neuropsychiatric disorders." *Nat Neurosci* 13 (10): 1161–1169. doi:10.1038/nn.2647.

Newmeyer, M. N., M. J. Swortwood, A. J. Barnes, O. A. Abulseoud, K. B. Scheidweiler, and M. A. Huestis. 2016. "Free and glucuronide whole blood cannabinoids' pharmacokinetics after controlled smoked, vaporized, and oral cannabis administration in frequent and occasional cannabis users: Identification of recent cannabis intake." *Clin Chem* 62 (12): 1579–1592. doi:10.1373/clinchem.2016.263475.

Nguyen, L. C., D. Yang, V. Nicolaescu, T. J. Best, H. Gula, D. Saxena, J. D. Gabbard, S.-N. Chen, T. Ohtsuki, J. B. Friesen, N. Drayman, A. Mohamed, C. Dann, D. Silva, L. Robinson-Mailman, A. Valdespino, L. Stock, E. Suarez, K. A. Jones, S.-A. Azizi, J. K. Demarco, W. E. Severson, C. D. Anderson, J. M. Millis, B. C. Dickinson, S. Tay, S. A. Oakes, G. F. Pauli, K. E. Palmer, The Natonal COVID Cohort Collaborative Consortium, D. O. Meltzer, G. Randall, and M. R. Rosner. 2022. "Cannabidiol inhibits SARS-CoV-2 replication through induction of the host ER stress and innate immune responses." *Science Advances*: eabi6110. doi:10.1126/sciadv.abi6110.

Nichols, J. M., and B. L. F. Kaplan. 2020. "Immune responses regulated by cannabidiol." *Cannabis Cannabinoid Res* 5: 12–31. doi:10.1089/can.2018.0073.

Nicholson, A. N., C. Turner, B. M. Stone, and P. J. Robson. 2004. "Effect of delta-9-tetrahydrocannabinol and cannabidiol on nocturnal sleep and early-morning behavior in young adults." *J Clin Psychopharmacol* 24 (3): 305–313. doi:10.1097/01.jcp.0000125688.05091.8f.

Nickles, M. A., and P. A. Lio. 2020. "Cannabinoids in dermatology: Hope or hype?" *Cannabis and Cannabinoid Research* 5 (4): 279–282. doi:10.1089/can.2019.0097.

Nitecka-Buchta, A., A. Nowak-Wachol, K. Wachol, K. Walczynska-Dragon, P. Olczyk, O. Batoryna, W. Kempa, and S. Baron. 2019. "Myorelaxant effect of transdermal cannabidiol application in patients with TMD: A randomized, double-blind trial." *J Clin Med* 8 (11): 1886. doi:10.3390/jcm8111886.

Nona, C. N., C. S. Hendershot, and B. Le Foll. 2019. "Effects of cannabidiol on alcohol-related outcomes: A review of preclinical and human research." *Exp Clin Psychopharmacol* 27 (4): 359–369. doi:10.1037/pha0000272.

Norrbrink Budh, C., I. Lund, P. Ertzgaard, A. Holtz, C. Hultling, R. Levi, L. Werhagen, and T. Lundeberg. 2003. "Pain in a Swedish spinal cord injury population." *Clin Rehabil* 17 (6): 685–690. doi:10.1191/0269215503cr664oa.

Norris, C., M. Loureiro, C. Kramar, J. Zunder, J. Renard, W. Rushlow, and S. R. Laviolette. 2016. "Cannabidiol modulates fear memory formation through interactions with serotonergic transmission in the mesolimbic system." *Neuropsychopharmacology* 41 (12): 2839–2850. doi:10.1038/npp.2016.93.

Notcutt, W., R. Langford, P. Davies, S. Ratcliffe, and R. Potts. 2012. "A placebo-controlled, parallel-group, randomized withdrawal study of subjects with symptoms of spasticity due to multiple sclerosis who are receiving long-term Sativex (nabiximols)." *Mult Scler* 18 (2): 219–228. doi:10.1177/1352458511419700.

Novotna, A., J. Mares, S. Ratcliffe, I. Novakova, M. Vachova, O. Zapletalova, C. Gasperini, C. Pozzilli, L. Cefaro, G. Comi, P. Rossi, Z. Ambler, Z. Stelmasiak, A. Erdmann, X. Montalban, A. Klimek, P. Davies, and Group Sativex Spasticity Study. 2011. "A randomized, double-blind, placebo-controlled, parallel-group, enriched-design study of nabiximols (Sativex), as add-on therapy, in subjects with refractory spasticity caused by multiple sclerosis." *Eur J Neurol* 18 (9): 1122–1131. doi:10.1111/j.1468-1331.2010.03328.x.

Nurmikko, T. J., M. G. Serpell, B. Hoggart, P. J. Toomey, B. J. Morlion, and D. Haines. 2007. "Sativex successfully treats neuropathic pain characterised by allodynia: A randomised, double-blind, placebo-controlled clinical trial." *Pain* 133 (1–3): 210–220. doi:10.1016/j.pain.2007.08.028.

O'Brien, L. D., C. L. Limebeer, E. M. Rock, G. Bottegoni, D. Piomelli, and L. A. Parker. 2013. "Anandamide transport inhibition by ARN272 attenuates nausea-induced behaviour in rats, and vomiting in shrews (*Suncus murinus*)." *Br J Pharmacol* 170 (5): 1130–1136. doi:10.1111/bph.12360.

O'Brien, L. D., K. L. Wills, B. Segsworth, B. Dashney, E. M. Rock, C. L. Limebeer, and L. A. Parker. 2013. "Effect of chronic exposure to rimonabant and phytocannabinoids on anxiety-like behavior and saccharin palatability." *Pharmacol Biochem Behav* 103 (3): 597–602. doi:10.1016/j.pbb.2012.10.008.

O'Connell, M., M. Sandgren, L. Frantzen, E. Bower, and B. Erickson. 2019. "Medical cannabis: Effects on opioid and benzodiazepine requirements for pain control." *Ann Pharmacother* 53 (11): 1081–1086. doi:10.1177/1060028019854221.

O'Neill, A., R. Wilson, G. Blest-Hopley, L. Annibale, M. Colizzi, M. Brammer, V. Giampietro, and S. Bhattacharyya. 2020. "Normalization of mediotemporal and prefrontal activity, and mediotemporal-striatal connectivity, may underlie antipsychotic effects of cannabidiol in psychosis." *Psychol Med* 51 (4): 596–606. doi:10.1017/S0033291719003519.

O'Shaughnessy, W. B. 1839. "On the prepartions of Indian hemp, or gunja (*Cannabis indica*); Their effects on the animal system in health, and their utility in the treatment of tentanus and other convulsive diseases." *Br Foreign Med Rev.* 10 (19): 225–228.

Ohlsson, A., J. E. Lindgren, S. Andersson, S. Agurell, H. Gillespie, and L. E. Hollister. 1986. "Single-dose kinetics of deuterium-labelled cannabidiol in man after smoking and intravenous administration." *Biomed Environ Mass Spectrom* 13 (2): 77–83. doi:10.1002/bms.1200130206.

Olah, A., B. I. Toth, I. Borbiro, K. Sugawara, A. G. Szollosi, G. Czifra, B. Pal, L. Ambrus, J. Kloepper, E. Camera, M. Ludovici, M. Picardo, T. Voets, C. C. Zouboulis, R. Paus, and T. Biro. 2014. "Cannabidiol exerts sebostatic and antiinflammatory effects on human sebocytes." *J Clin Invest* 124 (9): 3713–3724. doi:10.1172/JCI64628.

Onaivi, E. S., M. R. Green, and B. R. Martin. 1990. "Pharmacological characterization of cannabinoids in the elevated plus maze." *Journal of Pharmacology and Experimental Therapeutics* 253 (3): 1002–1009.

Orzalli, M. H., and J. C. Kagan. 2017. "Apoptosis and necroptosis as host defense strategies to prevent viral infection." *Trends Cell Biol* 27 (11): 800–809. doi:10.1016/j.tcb.2017.05.007.

Osborne, A. L., N. Solowij, and K. Weston-Green. 2017. "A systematic review of the effect of cannabidiol on cognitive function: Relevance to schizophrenia." *Neurosci Biobehav Rev* 72: 310–324. doi:10.1016/j.neubiorev.2016.11.012.

Oviedo, A., J. Glowa, and M. Herkenham. 1993. "Chronic cannabinoid administration alters cannabinoid receptor binding in rat brain: A quantitative autoradiographic study." *Brain Res* 616 (1–2): 293–302. doi:10.1016/0006-8993(93)90220-h.

Owen, M. J., M. C. O'Donovan, A. Thapar, and N. Craddock. 2011. "Neurodevelopmental hypothesis of schizophrenia." *Br J Psychiatry* 198 (3): 173–175. doi:10.1192/bjp.bp.110.084384.

Pacher, P., N. M. Kogan, and R. Mechoulam. 2020. "Beyond THC and endocannabinoids." *Annu Rev Pharmacol Toxicol* 60: 637–659. doi:10.1146/annurev-pharmtox-010818-021441.

Palmieri, B., C. Laurino, and M. Vadala. 2017. "Short-term efficacy of CBD-enriched hemp oil in girls with dysautonomic syndrome after human papillomavirus vaccination." *Isr Med Assoc J* 19 (2): 79–84.

Pan, H., P. Mukhopadhyay, M. Rajesh, V. Patel, B. Mukhopadhyay, B. Gao, G. Hasko, and P. Pacher. 2009. "Cannabidiol attenuates cisplatin-induced nephrotoxicity by decreasing oxidative/nitrosative stress, inflammation, and cell death." *J Pharmacol Exp Ther* 328 (3): 708–714. doi:10.1124/jpet.108.147181.

Panozzo, S., B. L. A. Collins, J. Weil, J. Whyte, M. Barton, M. Coperchini, M. Rametta, and J. Philip. 2020. "Who is asking about medicinal cannabis in palliative care?" *Internal Medicine Journal* 50 (2): 243–246. doi:10.1111/imj.14732.

Paolicelli, D., V. Direnzo, A. Manni, M. D'Onghia, C. Tortorella, S. Zoccolella, V. Di Lecce, A. Iaffaldano, and M. Trojano. 2016. "Long-term data of efficacy, safety, and tolerability in a real-life setting of THC/CBD oromucosal spray-treated multiple sclerosis patients." *J Clin Pharmacol* 56 (7): 845–851. doi:10.1002/jcph.670.

Paria, B. C., S. K. Das, and S. K. Dey. 1995. "The preimplantation mouse embryo is a target for cannabinoid ligand-receptor signaling." *Proc Natl Acad Sci USA* 92 (21): 9460–9464. doi:10.1073/pnas.92.21.9460.

Parker, L. A. 2014. "Conditioned flavor avoidance and conditioned gaping: rat models of conditioned nausea." *Eur J Pharmacol* 722: 122–133. doi:10.1016/j.ejphar.2013.09.070.

Parker, L. A., P. Burton, R. E. Sorge, C. Yakiwchuk, and R. Mechoulam. 2004. "Effect of low doses of delta9-tetrahydrocannabinol and cannabidiol on the extinction of cocaine-induced and amphetamine-induced conditioned place preference learning in rats." *Psychopharmacology (Berl)* 175 (3): 360–366. doi:10.1007/s00213-004-1825-7.

Parker, L. A., M. Kwiatkowska, P. Burton, and R. Mechoulam. 2004. "Effect of cannabinoids on lithium-induced vomiting in the *Suncus murinus* (house musk shrew)." *Psychopharmacology (Berl)* 171 (2): 156–161. doi:10.1007/s00213-003-1571-2.

Parker, L. A., and R. Mechoulam. 2003. "Cannabinoid agonists and antagonists modulate lithium-induced conditioned gaping in rats." *Integr Physiol Behav Sci.* 38 (2): 133–145.

Parker, L. A., R. Mechoulam, and C. Schlievert. 2002. "Cannabidiol, a nonpsychoactive component of cannabis and its synthetic dimethylheptyl homolog suppress nausea in an experimental model with rats." *Neuroreport* 13 (5): 567–570. doi:10.1097/00001756-200204160-00006.

Patrician, A., M. Versic-Bratincevic, T. Mijacika, I. Banic, M. Marendic, D. Sutlovic, Z. Dujic, and P. N. Ainslie. 2019. "Examination of a new delivery approach for oral cannabidiol in healthy subjects: A randomized, double-blinded, placebo-controlled pharmacokinetics study." *Adv Ther* 36 (11): 3196–3210. doi:10.1007/s12325-019-01074-6.

Patti, F., S. Messina, C. Solaro, M. P. Amato, R. Bergamaschi, S. Bonavita, R. Bruno Bossio, V. Brescia Morra, G. F. Costantino, P. Cavalla, D. Centonze, G. Comi, S. Cottone, M. Danni, A. Francia, A. Gajofatto, C. Gasperini, A. Ghezzi, A. Iudice, G. Lus, G. T. Maniscalco, M. G. Marrosu, M. Matta, M. Mirabella, E. Montanari, C. Pozzilli, M. Rovaris, E. Sessa, D. Spitaleri, M. Trojano, P. Valentino, M. Zappia, and Sa Fe Study Group. 2016. "Efficacy and safety of cannabinoid oromucosal spray for multiple sclerosis spasticity." *J Neurol Neurosurg Psychiatry* 87 (9): 944–951. doi:10.1136/jnnp-2015-312591.

Paudel, K. S., D. C. Hammell, R. U. Agu, S. Valiveti, and A. L. Stinchcomb. 2010. "Cannabidiol bioavailability after nasal and transdermal application: Effect of permeation enhancers." *Drug Dev Ind Pharm* 36 (9): 1088–1097. doi:10.3109/03639041003657295.

Pazos, M. R., V. Cinquina, A. Gomez, R. Layunta, M. Santos, J. Fernandez-Ruiz, and J. Martinez-Orgado. 2012. "Cannabidiol administration after hypoxia-ischemia to newborn rats reduces long-term brain injury and restores neurobehavioral function." *Neuropharmacology* 63 (5): 776–783. doi:10.1016/j.neuropharm.2012.05.034.

Pazos, M. R., N. Mohammed, H. Lafuente, M. Santos, E. Martinez-Pinilla, E. Moreno, E. Valdizan, J. Romero, A. Pazos, R. Franco, C. J. Hillard, F. J. Alvarez, and J. Martinez-Orgado. 2013. "Mechanisms of cannabidiol neuroprotection in hypoxic-ischemic newborn pigs: Role of 5HT(1A) and CB2 receptors." *Neuropharmacology* 71: 282–291. doi:10.1016/j.neuropharm.2013.03.027.

Pedrazzi, J. F., A. C. Issy, F. V. Gomes, F. S. Guimaraes, and E. A. Del-Bel. 2015. "Cannabidiol effects in the prepulse inhibition disruption induced by amphetamine." *Psychopharmacology (Berl)* 232 (16): 3057–3065. doi:10.1007/s00213-015-3945-7.

Pellow, S., P. Chopin, S. E. File, and M. Briley. 1985. "Validation of open:closed arm entries in an elevated plus-maze as a measure of anxiety in the rat." *J Neurosci Methods* 14 (3): 149–167. doi:10.1016/0165-0270(85)90031-7.

Peres, F. F., M. C. Diana, R. Levin, M. A. Suiama, V. Almeida, A. M. Vendramini, C. M. Santos, A. W. Zuardi, J. E. C. Hallak, J. A. Crippa, and V. C. Abilio. 2018. "Cannabidiol administered during peri-adolescence prevents behavioral abnormalities in an animal model of schizophrenia." *Front Pharmacol* 9: 901. doi:10.3389/fphar.2018.00901.

Pertwee, R. G. 2008. "The diverse CB1 and CB2 receptor pharmacology of three plant cannabinoids: delta9-tetrahydrocannabinol, cannabidiol and delta9-tetrahydrocannabivarin." *Br J Pharmacol* 153 (2): 199–215. doi:10.1038/sj.bjp.0707442.

Pertwee, R. G., E. M. Rock, K. Guenther, C. L. Limebeer, L. A. Stevenson, C. Haj, R. Smoum, L. A. Parker, and R. Mechoulam. 2018. "Cannabidiolic acid methyl ester, a stable synthetic analogue of cannabidiolic acid, can produce 5-HT." *Br J Pharmacol* 175 (1): 100–112. doi:10.1111/bph.14073.

Petrzilka, T. W., W. Haefliger, c. Sikemeir, G. Ohloff, and A. Eschenmoser. 1967. "Synthese und chiralitat des (–) cannabidiols." *Helv Chim Acta* 50: 719–723.

Philpott, H. T., M. O'Brien, and J. J. McDougall. 2017. "Attenuation of early phase inflammation by cannabidiol prevents pain and nerve damage in rat osteoarthritis." *Pain* 158 (12): 2442–2451. doi:10.1097/j.pain.0000000000001052.

Pinto, J. V., G. Saraf, C. Frysch, D. Vigo, K. Keramatian, T. Chakrabarty, R. W. Lam, M. Kauer-Sant'Anna, and L. N. Yatham. 2020. "Cannabidiol as a treatment for mood disorders: A systematic review." *Can J Psychiatry* 65 (4): 213–227. doi:10.1177/0706743719895195.

Piper, B. J. 2018. "Mother of berries, ACDC, or chocolope: Examination of the strains used by medical cannabis patients in New England." *J Psychoactive Drugs* 50 (2): 95–104. doi:10.1080/02791072.2017.1390179.

Pisanti, S., A. M. Malfitano, E. Ciaglia, A. Lamberti, R. Ranieri, G. Cuomo, M. Abate, G. Faggiana, M. C. Proto, D. Fiore, C. Laezza, and M. Bifulco. 2017. "Cannabidiol: State of the art and new challenges for therapeutic applications." *Pharmacol Ther* 175: 133–150. doi:10.1016/j.pharmthera.2017.02.041.

Poli-Bigelli, S., J. Rodrigues-Pereira, A. D. Carides, G. Julie Ma, K. Eldridge, A. Hipple, J. K. Evans, K. J. Horgan, F. Lawson, and Group Aprepitant Protocol 054 Study. 2003. "Addition of the neurokinin 1 receptor antagonist aprepitant to standard antiemetic therapy improves control of chemotherapy-induced nausea and vomiting. Results from a randomized, double-blind, placebo-controlled trial in Latin America." *Cancer* 97 (12): 3090–3098. doi:10.1002/cncr.11433.

Porsolt, R. D., G. Anton, N. Blavet, and M. Jalfre. 1978. "Behavioural despair in rats: A new model sensitive to antidepressant treatments." *Eur J Pharmacol* 47 (4): 379–391. doi:10.1016/0014-2999(78)90118-8.

Porsolt, R. D., M. Le Pichon, and M. Jalfre. 1977. "Depression: A new animal model sensitive to antidepressant treatments." *Nature* 266 (5604): 730–732. doi:10.1038/266730a0.

Portenoy, R. K., E. D. Ganae-Motan, S. Allende, R. Yanagihara, L. Shaiova, S. Weinstein, R. McQuade, S. Wright, and M. T. Fallon. 2012. "Nabiximols for opioid-treated cancer patients with poorly-controlled chronic pain: A randomized, placebo-controlled, graded-dose trial." *J Pain* 13 (5): 438–449. doi:10.1016/j.jpain.2012.01.003.

Porter, B. E., and C. Jacobson. 2013. "Report of a parent survey of cannabidiol-enriched cannabis use in pediatric treatment-resistant epilepsy." *Epilepsy Behav* 29 (3): 574–577. doi:10.1016/j.yebeh.2013.08.037.

Press, C. A., K. G. Knupp, and K. E. Chapman. 2015. "Parental reporting of response to oral cannabis extracts for treatment of refractory epilepsy." *Epilepsy Behav* 45: 49–52. doi:10.1016/j.yebeh.2015.02.043.

Pryce, G., and D. Baker. 2014. "Cannabis and multiple sclerosis." In *Handbook of Cannabis*, edited by Roger G. Pertwee, 487–501. Oxford: Oxford University Press.

Pucci, M., C. Rapino, A. Di Francesco, E. Dainese, C. D'Addario, and M. Maccarrone. 2013. "Epigenetic control of skin differentiation genes by phytocannabinoids." *Br J Pharmacol* 170 (3): 581–591. doi:10.1111/bph.12309.

Qian, Y., T. K. Gilliland, and J. S. Markowitz. 2020. "The influence of carboxylesterase 1 polymorphism and cannabidiol on the hepatic metabolism of heroin." *Chem Biol Interact* 316: 108914. doi:10.1016/j.cbi.2019.108914.

Raj, V., J. G. Park, K. H. Cho, P. Choi, T. Kim, J. Ham, and J. Lee. 2021. "Assessment of antiviral potencies of cannabinoids against SARS-CoV-2 using computational and in vitro approaches." *Int J Biol Macromol* 168: 474–485. doi:10.1016/j.ijbiomac.2020.12.020.

Rajesh, M., P. Mukhopadhyay, S. Batkai, V. Patel, K. Saito, S. Matsumoto, Y. Kashiwaya, B. Horvath, B. Mukhopadhyay, L. Becker, G. Hasko, L. Liaudet, D. A. Wink, A. Veves, R. Mechoulam, and P. Pacher. 2010. "Cannabidiol attenuates cardiac dysfunction, oxidative stress, fibrosis, and inflammatory and cell death signaling pathways in diabetic cardiomyopathy." *J Am Coll Cardiol* 56 (25): 2115–2125. doi:10.1016/j.jacc.2010.07.033.

Ramer, R., and B. Hinz. 2017. "Cannabinoids as anticancer drugs." *Adv Pharmacol* 80: 397–436. doi:10.1016/bs.apha.2017.04.002.

Raymundi, A. M., T. R. da Silva, A. R. Zampronio, F. S. Guimaraes, L. J. Bertoglio, and C. A. J. Stern. 2020. "A time-dependent contribution of hippocampal CB1, CB2 and PPAR gamma receptors to cannabidiol-induced disruption of fear memory consolidation." *Br J Pharmacol* 177 (4): 945–957. doi:10.1111/bph.14895.

RECOVERY Collaborative Group. 2021. "Dexamethasone in hospitalized patients with Covid-19." *New England Journal of Medicine* 384 (8): 693–704. doi:10.1056/NEJMoa2021436

Reilly, D., P. Didcott, W. Swift, and W. Hall. 1998. "Long-term cannabis use: Characteristics of users in an Australian rural area." *Addiction* 93 (6): 837–846. doi:10.1046/j.1360-0443.1998.9368375.x.

Reiss, C. S. 2010. "Cannabinoids and viral infections." *Pharmaceuticals (Basel)* 3 (6): 1873–1886. doi:10.3390/ph3061873.

Ren, Y., J. Whittard, A. Higuera-Matas, C. V. Morris, and Y. L. Hurd. 2009. "Cannabidiol, a nonpsychotropic component of cannabis, inhibits cue-induced heroin seeking and normalizes discrete mesolimbic neuronal disturbances." *J Neurosci* 29 (47): 14764–14769. doi:10.1523/JNEUROSCI.4291-09.2009.

Renard, J., M. Loureiro, L. G. Rosen, J. Zunder, C. de Oliveira, S. Schmid, W. J. Rushlow, and S. R. Laviolette. 2016. "Cannabidiol counteracts amphetamine-induced neuronal and behavioral sensitization of the mesolimbic dopamine pathway through a novel mTOR/p70S6 kinase signaling pathway." *J Neurosci* 36 (18): 5160–5169. doi:10.1523/JNEUROSCI.3387-15.2016.

Renard, J., C. Norris, W. Rushlow, and S. R. Laviolette. 2017. "Neuronal and molecular effects of cannabidiol on the mesolimbic dopamine system: Implications for novel schizophrenia treatments." *Neurosci Biobehav Rev* 75: 157–165. doi:10.1016/j.neubiorev.2017.02.006.

Resstel, L. B., S. R. Joca, F. A. Moreira, F. M. Correa, and F. S. Guimaraes. 2006. "Effects of cannabidiol and diazepam on behavioral and cardiovascular responses induced by contextual conditioned fear in rats." *Behav Brain Res* 172 (2): 294–298. doi:10.1016/j.bbr.2006.05.016.

Resstel, L. B., R. F. Tavares, S. F. Lisboa, S. R. Joca, F. M. Correa, and F. S. Guimaraes. 2009. "5-HT1A receptors are involved in the cannabidiol-induced attenuation of behavioural and cardiovascular responses to acute restraint stress in rats." *Br J Pharmacol* 156 (1): 181–188. doi:10.1111/j.1476-5381.2008.00046.x.

Reus, G. Z., R. B. Stringari, K. F. Ribeiro, T. Luft, H. M. Abelaira, G. R. Fries, B. W. Aguiar, F. Kapczinski, J. E. Hallak, A. W. Zuardi, J. A. Crippa, and J. Quevedo. 2011. "Administration of cannabidiol and imipramine induces antidepressant-like effects in the forced swimming test and increases brain-derived neurotrophic factor levels in the rat amygdala." *Acta Neuropsychiatr* 23 (5): 241–248. doi:10.1111/j.1601-5215.2011.00579.x.

Reynolds, J. R. 1868. "Therapeutical uses and toxic effects of *Cannabis indica*." *Lancet* 1:637–638.

Ribeiro, A., V. I. Almeida, C. Costola-de-Souza, V. Ferraz-de-Paula, M. L. Pinheiro, L. B. Vitoretti, J. Gimenes-Junior, A. T. Akamine, J. A. Crippa, W. Tavares-de-Lima, and J. Palermo-Neto. 2015. "Cannabidiol improves lung function and inflammation in mice submitted to LPS-induced acute lung injury." *Immunopharmacology and Immunotoxicology* 37 (1): 35–41. doi: 10.3109/08923973.2014.976794

Ribeiro, A., V. Ferraz-de-Paula, M. L. Pinheiro, L. B. Vitoretti, D. P. Mariano-Souza, W. M. Quinteiro-Filho, A. T. Akamine, V. I. Almeida, J. Quevedo, F. Dal-Pizzol, J. E. Hallak, A. W. Zuardi, J. A. Crippa, and J. Palermo-Neto. 2012. "Cannabidiol, a non-psychotropic plant-derived cannabinoid, decreases inflammation in a murine model of acute lung injury: Role for the adenosine A(2A) receptor." *Eur J Pharmacol* 678 (1–3): 78–85. doi:10.1016/j.ejphar.2011.12.043.

Richards, J. R. 2017. "Cannabinoid hyperemesis syndrome: A disorder of the HPA axis and sympathetic nervous system?" *Med Hypotheses* 103: 90–95. doi:10.1016/j.mehy.2017.04.018.

Riedel, G., P. Fadda, S. McKillop-Smith, R. G. Pertwee, B. Platt, and L. Robinson. 2009. "Synthetic and plant-derived cannabinoid receptor antagonists show hypophagic properties in fasted and non-fasted mice." *Br J Pharmacol* 156 (7): 1154–1166. doi:10.1111/j.1476-5381.2008.00107.x.

Rock, E. M., D. Bolognini, C. L. Limebeer, M. G. Cascio, S. Anavi-Goffer, P.J. Fletcher, R. Mechoulam, R. G. Pertwee, and L. A. Parker. 2012. "Cannabidiol, a non-psychotropic component of cannabis, attenuates vomiting and nausea-like behaviour via indirect agonism of 5-HT(1A) somatodendritic autoreceptors in the dorsal raphe nucleus." *British Journal of Pharmacology* 165 (8): 2620–2634. doi:10.1111/j.1476-5381.2011.01621.x.

Rock, E. M., C. Connolly, C. L. Limebeer, and L. A. Parker. 2016. "Effect of combined oral doses of Δ(9)-tetrahydrocannabinol (THC) and cannabidiolic acid (CBDA) on acute and anticipatory nausea in rat models." *Psychopharmacology (Berl)* 233 (18): 3353–3360. doi:10.1007/s00213-016-4378-7.

Rock, E. M., J. M. Goodwin, C. L. Limebeer, A. Breuer, R. G. Pertwee, R. Mechoulam, and L. A. Parker. 2011. "Interaction between non-psychotropic cannabinoids in marihuana: Effect of cannabigerol (CBG) on the antinausea or anti-emetic effects of cannabidiol (CBD) in rats and shrews." *Psychopharmacology (Berl)* 215 (3): 505–512. doi:10.1007/s00213-010-2157-4.

Rock, E. M., C. L. Limebeer, R. Mechoulam, D. Piomelli, and L. A. Parker. 2008. "The effect of cannabidiol and URB597 on conditioned gaping (a model of nausea) elicited by a lithium-paired context in the rat." *Psychopharmacology (Berl)* 196 (3): 389–395. doi:10.1007/s00213-007-0970-1.

Rock, E. M., C. L. Limebeer, R. Navaratnam, M. A. Sticht, N. Bonner, K. Engeland, R. Downey, H. Morris, M. Jackson, and L. A. Parker. 2014. "A comparison of cannabidiolic acid with other treatments for anticipatory nausea using a rat model of contextually elicited conditioned gaping." *Psychopharmacology (Berl)* 231 (16): 3207–3215. doi:10.1007/s00213-014-3498-1.

Rock, E. M., C. L. Limebeer, and L. A. Parker. 2015. "Effect of combined doses of $\Delta(9)$-tetrahydrocannabinol (THC) and cannabidiolic acid (CBDA) on acute and anticipatory nausea using rat (Sprague-Dawley) models of conditioned gaping." *Psychopharmacology (Berl)* 232 (24): 4445–4454. doi:10.1007/s00213-015-4080-1.

Rock, E. M., C. L. Limebeer, and L. A. Parker. 2018. "Effect of cannabidiolic acid and (9)-tetrahydrocannabinol on carrageenan-induced hyperalgesia and edema in a rodent model of inflammatory pain." *Psychopharmacology (Berl)* 235 (11): 3259–3271. doi:10.1007/s00213-018-5034-1.

Rock, E. M., C. L. Limebeer, G. N. Petrie, L. A. Williams, R. Mechoulam, and L. A. Parker. 2017. "Effect of prior foot shock stress and delta(9)-tetrahydrocannabinol, cannabidiolic acid, and cannabidiol on anxiety-like responding in the light-dark emergence test in rats." *Psychopharmacology (Berl)* 234 (14): 2207–2217. doi:10.1007/s00213-017-4626-5.

Rock, E. M., and L. A. Parker. 2013. "Effect of low doses of cannabidiolic acid and ondansetron on LiCl-induced conditioned gaping (a model of nausea-induced behaviour) in rats." *Br J Pharmacol.* 169 (3): 685–692. doi:10.1111/bph.12162.

Rock, E. M., and L. A. Parker. 2015. "Synergy between cannabidiol, cannabidiolic acid, and delta(9)-tetrahydrocannabinol in the regulation of emesis in the *Suncus murinus* (house musk shrew)." *Behav Neurosci* 129 (3): 368–370. doi:10.1037/bne0000057.

Rock, E. M., M. T. Sullivan, S. A. Collins, H. Goodman, C. L. Limebeer, R. Mechoulam, and L. A. Parker. 2020. "Evaluation of repeated or acute treatment with cannabidiol (CBD), cannabidiolic acid (CBDA) or CBDA methyl ester (HU-580) on nausea and/or vomiting in rats and shrews." *Psychopharmacology (Berl)* 237 (9): 2621–2631. doi:10.1007/s00213-020-05559-z.

Rock, E. M., M. T. Sullivan, S. Pravato, M. Pratt, C. L. Limebeer, and L. A. Parker. 2020. "Effect of combined doses of delta(9)-tetrahydrocannabinol and cannabidiol or tetrahydrocannabinolic acid and cannabidiolic acid on acute nausea in male Sprague-Dawley rats." *Psychopharmacology (Berl)* 237 (3): 901–914. doi:10.1007/s00213-019-05428-4.

Rodrigues da Silva, N., F. V. Gomes, A. B. Sonego, and F. S. Guimaraes. 2020. "Cannabidiol attenuates behavioral changes in a rodent model of schizophrenia through 5-HT1A, but not CB1 and CB2 receptors." *Pharmacol Res* 156: 104749. doi:10.1016/j.phrs.2020.104749.

Rodriguez-Munoz, M., Y. Onetti, E. Cortes-Montero, J. Garzon, and P. Sanchez-Blazquez. 2018. "Cannabidiol enhances morphine antinociception, diminishes NMDA-mediated seizures and reduces stroke damage via the sigma 1 receptor." *Mol Brain* 11 (1): 51. doi:10.1186/s13041-018-0395-2.

Rog, D. J., T. J. Nurmikko, T. Friede, and C. A. Young. 2005. "Randomized, controlled trial of cannabis-based medicine in central pain in multiple sclerosis." *Neurology* 65 (6): 812–819. doi:10.1212/01.wnl.0000176753.45410.8b.

Romero, J., F. Berrendero, J. Manzanares, A. Pérez, J. Corchero, J. A. Fuentes, J. J. Fernández-Ruiz, and J. A. Ramos. 1998. "Time-course of the cannabinoid receptor down-regulation in the adult rat brain caused by repeated exposure to delta9-tetrahydrocannabinol." *Synapse* 30 (3): 298–308. doi:10.1002/(sici)1098-2396(199811)30:3<298::aid-syn7>3.0.co;2–6.

Rosenkrantz, H., R. W. Fleischman, and R. J. Grant. 1981. "Toxicity of short-term administration of cannabinoids to rhesus monkeys." *Toxicol Appl Pharmacol* 58 (1): 118–131. doi:10.1016/0041-008x(81)90122-8.

Rosenthal, M. S. 1972. "Clinical effects of marijuana on the young: A call for more systematic clinical inquiry." *Int J Psychiatry* 10 (2): 75–77.

Roser, P., J. Gallinat, G. Weinberg, G. Juckel, I. Gorynia, and A. M. Stadelmann. 2009. "Psychomotor performance in relation to acute oral administration of delta9-tetrahydrocannabinol and standardized cannabis extract in healthy human subjects." *Eur Arch Psychiatry Clin Neurosci* 259 (5): 284–292. doi:10.1007/s00406-009-0868-5.

Ross, D. A., M. R. Arbuckle, M. J. Travis, J. B. Dwyer, G. I. van Schalkwyk, and K. J. Ressler. 2017. "An integrated neuroscience perspective on formulation and treatment planning for posttraumatic stress disorder: An educational review." *JAMA Psychiatry* 74 (4): 407–415. doi:10.1001/jamapsychiatry.2016.3325.

Rossignoli, M. T., C. Lopes-Aguiar, R. N. Ruggiero, R. A. Do Val da Silva, L. S. Bueno-Junior, L. Kandratavicius, J. E. Peixoto-Santos, J. A. Crippa, J. E. Cecilio Hallak, A. W. Zuardi, R. E. Szawka, J. Anselmo-Franci, J. P. Leite, and R. N. Romcy-Pereira. 2017. "Selective post-training time window for memory consolidation interference of cannabidiol into the prefrontal cortex: Reduced dopaminergic modulation and immediate gene expression in limbic circuits." *Neuroscience* 350: 85–93. doi:10.1016/j.neuroscience.2017.03.019.

Ruan, Q., K. Yang, W. Wang, L. Jiang, and J. Song. 2020. "Clinical predictors of mortality due to COVID-19 based on an analysis of data of 150 patients from Wuhan, China." *Intensive Care Medicine* 46 (5): 846–848. doi:10.1007/s00134-020-05991-x.

Rubino, T., L. Tizzoni, D. Viganò, P. Massi, and D. Parolaro. 1997. "Modulation of rat brain cannabinoid receptors after chronic morphine treatment." *Neuroreport* 8 (15): 3219–3223. doi:10.1097/00001756-199710200-00007.

Rudd, J. A., E. Nalivaiko, N. Matsuki, C. Wan, and P. L. R. Andrews. 2015. "The involvement of TRPV1 in emesis and anti-emesis." *Temperature* 2 (2): 258–276. doi:10.1080/23328940.2015.1043042.

Runner, R. P., A. N. Luu, N. A. Nassif, T. S. Scudday, J. J. Patel, S. L. Barnett, and R. S. Gorab. 2020. "Use of tetrahydrocannabinol and cannabidiol products in the perioperative period around primary unilateral total hip and knee arthroplasty." *J Arthroplasty* 35 (6S): S138–S143. doi:10.1016/j.arth.2020.01.077.

Russo, E. B. 2017. "Cannabidiol claims and misconceptions." *Trends Pharmacol Sci* 38 (3): 198–201. doi:10.1016/j.tips.2017.03.006.

Russo, E. B., A. Burnett, B. Hall, and K. K. Parker. 2005. "Agonistic properties of cannabidiol at 5-HT1a receptors." *Neurochem Res* 30 (8): 1037–1043. doi:10.1007/s11064-005-6978-1.

Russo, E. B., C. Spooner, L. May, R. Leslie, and V. L. Whiteley. 2021. "Cannabinoid hyperemesis syndrome survey and genomic investigation." Online ahead of print. *Cannabis Cannabinoid Res* doi:10.1089/can.2021.0046.

Russo, E. B., G. W. Guy, and P. J. Robson. 2007. "Cannabis, pain, and sleep: Lessons from therapeutic clinical trials of Sativex, a cannabis-based medicine." *Chem Biodivers* 4 (8): 1729–1743. doi:10.1002/cbdv.200790150.

Russo, E. B., A. Burnett, B. Hall, and Parker K. K. 2005. "Agonistic properties of cannabidiol at 5-HT1a receptors." *Neurochem Res* 30: 1037–1043. doi:10.1007/s11064-005-6978-1.

Ryberg, E., N. Larsson, S. Sjogren, S. Hjorth, N. O. Hermansson, J. Leonova, T. Elebring, K. Nilsson, T. Drmota, and P. J. Greasley. 2007. "The orphan receptor GPR55 is a novel cannabinoid receptor." *Br J Pharmacol* 152 (7): 1092–1101. doi:10.1038/sj.bjp.0707460.

Saft, C., S. M. von Hein, T. Lucke, C. Thiels, M. Peball, A. Djamshidian, B. Heim, and K. Seppi. 2018. "Cannabinoids for treatment of dystonia in Huntington's disease." *J Huntington's Dis* 7 (2): 167–173. doi:10.3233/JHD-170283.

Sales, A. J., C. C. Crestani, F. S. Guimaraes, and S. R. L. Joca. 2018. "Antidepressant-like effect induced by cannabidiol is dependent on brain serotonin levels." *Prog Neuropsychopharmacol Biol Psychiatry* 86: 255–261. doi:10.1016/j.pnpbp.2018.06.002.

Sales, A. J., M. V. Fogaca, A. G. Sartim, V. S. Pereira, G. Wegener, F. S. Guimaraes, and S. R. L. Joca. 2019. "Cannabidiol induces rapid and sustained antidepressant-like effects through increased BDNF signaling and synaptogenesis in the prefrontal cortex." *Mol Neurobiol* 56 (2): 1070–1081. doi:10.1007/s12035-018-1143-4.

Sales, A. J., F. S. Guimarães, and S. R. L. Joca. 2020. "CBD modulates DNA methylation in the prefrontal cortex and hippocampus of mice exposed to forced swim." *Behav Brain Res* 388: 112627. doi:10.1016/j.bbr.2020.112627.

Samara, E., M. Bialer, and R. Mechoulam. 1988. "Pharmacokinetics of cannabidiol in dogs." *Drug Metab Dispos* 16 (3): 469–472.

Sanmartin, P. E., and K. Detyniecki. 2018. "Cannabidiol for epilepsy: New hope on the horizon?" *Clin Ther* 40 (9): 1438–1441. doi:10.1016/j.clinthera.2018.07.020.

Santavy, F. 1964. "Notes on the structures of cannabidiol compounds." *Acta Univ Palackianae Olamuc* 35:5–6.

Scarante, F. F., M. A. Ribeiro, A. F. Almeida-Santos, F. S. Guimaraes, and A. C. Campos. 2020. "Glial cells and their contribution to the mechanisms of action of cannabidiol in neuropsychiatric disorders." *Front Pharmacol* 11:618065. doi:10.3389/fphar.2020.618065.

Scherma, M., L. V. Panlilio, P. Fadda, L. Fattore, I. Gamaleddin, B. Le Foll, Z. Justinová, E. Mikics, J. Haller, J. Medalie, J. Stroik, C. Barnes, S. Yasar, G. Tanda, D. Piomelli, W. Fratta, and S. R. Goldberg. 2008 "Inhibition of anandamide hydrolysis by cyclohexyl carbamic acid 3′-carbamoyl-3-yl ester (URB597) reverses abuse-related behavioral and neurochemical effects of nicotine in rats." *J Pharmacol Exp Ther* 327: 482–490. doi:10.1124/jpet.108.142224.

Schiavon, A. P., J. M. Bonato, H. Milani, F. S. Guimaraes, and R. M. Weffort de Oliveira. 2016. "Influence of single and repeated cannabidiol administration on emotional behavior and markers of cell proliferation and neurogenesis in non-stressed mice." *Prog Neuropsychopharmacol Biol Psychiatry* 64: 27–34. doi:10.1016/j.pnpbp.2015.06.017.

Schicho, R., and M. Storr. 2012. "Topical and systemic cannabidiol improves trinitrobenzene sulfonic acid colitis in mice." *Pharmacology* 89 (3–4): 149–155. doi:10.1159/000336871.

Schilling, J. M., C. G. Hughes, M. S. Wallace, M. Sexton, M. Backonja, and T. Moeller-Bertram. 2021. "Cannabidiol as a treatment for chronic pain: A survey of patients' perspectives and attitudes." *J Pain Res* 14: 1241–1250. doi:10.2147/jpr.s278718.

Schleicher, E. M., F. W. Ott, M. Muller, B. Silcher, M. E. Sichler, M. J. Low, J. M. Wagner, and Y. Bouter. 2019. "Prolonged cannabidiol treatment lacks on detrimental effects on memory, motor performance and anxiety in C57BL/6J mice." *Front Behav Neurosci* 13: 94. doi:10.3389/fnbeh.2019.00094.

Schneider, T., L. Zurbriggen, M. Dieterle, E. Mauermann, P. Frei, K. Mercer-Chalmers-Bender, and W. Ruppen. 2022. "Pain response to cannabidiol in induced acute nociceptive pain, allodynia, and hyperalgesia by using a model mimicking acute pain in healthy adults in a randomized trial (CANAB I)." *Pain* 163 (1): e62–e71. doi:10.1097/j.pain.0000000000002310.

Schoevers, J., J. E. Leweke, and F. M. Leweke. 2020. "Cannabidiol as a treatment option for schizophrenia: Recent evidence and current studies." *Curr Opin Psychiatry* 33 (3): 185–191. doi:10.1097/YCO.0000000000000596.

Schofield, D., C. Tennant, L. Nash, L. Degenhardt, A. Cornish, C. Hobbs, and G. Brennan. 2006. "Reasons for cannabis use in psychosis." *Aust NZ J Psychiatry* 40 (6–7): 570–574. doi:10.1080/j.1440-1614.2006.01840.x.

Schubart, C. D., I. E. Sommer, W. A. van Gastel, R. L. Goetgebuer, R. S. Kahn, and M. P. Boks. 2011. "Cannabis with high cannabidiol content is associated with fewer psychotic experiences." *Schizophr Res* 130 (1–3): 216–221. doi:10.1016/j.schres.2011.04.017.

Scopinho, A. A., F. S. Guimaraes, F. M. Correa, and L. B. Resstel. 2011. "Cannabidiol inhibits the hyperphagia induced by cannabinoid-1 or serotonin-1A receptor agonists." *Pharmacol Biochem Behav* 98 (2): 268–272. doi:10.1016/j.pbb.2011.01.007.

Serpell, M. G., W. Notcutt, and C. Collin. 2013. "Sativex long-term use: An open-label trial in patients with spasticity due to multiple sclerosis." *J Neurol* 260 (1): 285–295. doi:10.1007/s00415-012-6634-z.

Serpell, M., S. Ratcliffe, J. Hovorka, M. Schofield, L. Taylor, H. Lauder, and E. Ehler. 2014. "A double-blind, randomized, placebo-controlled, parallel group study of THC/CBD spray in peripheral neuropathic pain treatment." *Eur J Pain* 18 (7): 999–1012. doi:10.1002/j.1532-2149.2013.00445.x.

Shallcross, J., P. Hamor, A. R. Bechard, M. Romano, L. Knackstedt, and M. Schwendt. 2019. "The divergent effects of CDPPB and cannabidiol on fear extinction and anxiety in a predator scent stress model of PTSD in rats." *Front Behav Neurosci* 13: 91. doi:10.3389/fnbeh.2019.00091.

Shani, A., and R. Mechoulam. 1971. "Photochemical reactions of cannabidiol. Cyclization to 1-THC and other transformations." *Tetrahedron* 27: 601–606.

Shannon, S., N. Lewis, H. Lee, and S. Hughes. 2019. "Cannabidiol in anxiety and sleep: A large case series." *Perm J* 23: 18–41. doi:10.7812/TPP/18-041.

Shannon, S., and J. Opila-Lehman. 2015. "Cannabidiol oil for decreasing addictive use of marijuana: A case report." *Integr Med (Encinitas)* 14 (6): 31–35.

Shannon, S., and J. Opila-Lehman. 2016. "Effectiveness of cannabidiol oil for pediatric anxiety and insomnia as part of posttraumatic stress disorder: A case report." *Perm J* 20 (4): 16-005. doi:10.7812/TPP/16-005.

Shbiro, L., D. Hen-Shoval, N. Hazut, K. Rapps, S. Dar, G. Zalsman, R. Mechoulam, A. Weller, and G. Shoval. 2019. "Effects of cannabidiol in males and females in two different rat models of depression." *Physiol Behav* 201: 59–63. doi:10.1016/j.physbeh.2018.12.019.

Sholler, D. J., L. Schoene, and T. R. Spindle. 2020. "Therapeutic efficacy of cannabidiol (CBD): A review of the evidence from clinical trials and human laboratory studies." *Curr Addict Rep* 7 (3): 405–412. doi:10.1007/s40429-020-00326-8.

Shoval, G., L. Shbiro, L. Hershkovitz, N. Hazut, G. Zalsman, R. Mechoulam, and A. Weller. 2016. "Prohedonic effect of cannabidiol in a rat model of depression." *Neuropsychobiology* 73 (2): 123–129. doi:10.1159/000443890.

Shover, C. L., and K. Humphreys. 2020. "Debunking cannabidiol as a treatment for COVID-19: Time for the FDA to adopt a focused deterrence model?" *Cureus* 12 (6): e8671. doi:10.7759/cureus.8671.

Sihota, A., B. K. Smith, S. A. Ahmed, A. Bell, A. Blain, H. Clarke, Z. D. Cooper, C. Cyr, P. Daeninck, A. Deshpande, K. Ethans, D. Flusk, B. Le Foll, M. J. Milloy, D. E. Moulin, V. Naidoo, M. Ong, J. Perez, K. Rod, R. Sealey, D. Sulak, Z. Walsh, and C. O'Connell. 2020. "Consensus-based recommendations for titrating cannabinoids and tapering opioids for chronic pain control." *Int J Clin Pract* 75 (8): e13871. doi:10.1111/ijcp.13871.

Silva, N. R., F. V. Gomes, M. D. Fonseca, R. Mechoulam, A. Breuer, T. M. Cunha, and F. S. Guimaraes. 2017. "Antinociceptive effects of HUF-101, a fluorinated cannabidiol derivative." *Prog Neuropsychopharmacol Biol Psychiatry* 79 (Pt. B): 369–377. doi:10.1016/j.pnpbp.2017.07.012.

Silva, R. L., G. T. Silveira, C. W. Wanderlei, N. T. Cecilio, A. G. M. Maganin, M. Franchin, L. M. M. Marques, N. P. Lopes, J. A. Crippa, F. S. Guimaraes, J. C. F. Alves-Filho, F. Q. Cunha, and T. M. Cunha. 2019. "DMH-CBD, a cannabidiol analog with reduced cytotoxicity, inhibits TNF production by targeting NF-kB activity dependent on A2A receptor." *Toxicol Appl Pharmacol* 368: 63–71. doi:10.1016/j.taap.2019.02.011.

Sim-Selley, L. J. 2003. "Regulation of cannabinoid CB1 receptors in the central nervous system by chronic cannabinoids." *Crit Rev Neurobiol* 15 (2): 91–119. doi:10.1615/critrevneurobiol.v15.i2.10.

Simmerman, E., X. Qin, J. C. Yu, and B. Baban. 2019. "Cannabinoids as a potential new and novel treatment for melanoma: A pilot study in a murine model." *J Surg Res* 235: 210–215. doi:10.1016/j.jss.2018.08.055.

Singer, E., J. Judkins, N. Salomonis, L. Matlaf, P. Soteropoulos, S. McAllister, and L. Soroceanu. 2015. "Reactive oxygen species-mediated therapeutic response and resistance in glioblastoma." *Cell Death Dis* 6 (1): e1601. doi:10.1038/cddis.2014.566.

Sledzinski, P., J. Zeyland, R. Slomski, and A. Nowak. 2018. "The current state and future perspectives of cannabinoids in cancer biology." *Cancer Med* 7 (3): 765–775. doi:10.1002/cam4.1312.

Smith, P. B., S. P. Welch, and B. R. Martin. 1994. "Interactions between delta 9-tetrahydrocannabinol and kappa opioids in mice." *J Pharmacol Exp Ther* 268 (3): 1381–1387.

Sofia, R. D., and L. C. Knobloch. 1976. "Comparative effects of various naturally occurring cannabinoids on food, sucrose and water consumption by rats." *Pharmacol Biochem Behav* 4 (5): 591–599. doi:10.1016/0091-3057 (76)90202-1.

Sofia, R. D., H. B. Vassar, and L. C. Knobloch. 1975. "Comparative analgesic activity of various naturally occurring cannabinoids in mice and rats." *Psychopharmacologia* 40 (4): 285–295. doi:10.1007/bf00421466.

Solinas, M., P. Massi, V. Cinquina, M. Valenti, D. Bolognini, M. Gariboldi, E. Monti, T. Rubino, and D. Parolaro. 2013. "Cannabidiol, a non-psychoactive cannabinoid compound, inhibits proliferation and invasion in U87-MG and T98G glioma cells through a multitarget effect." *PLoS One* 8 (10): e76918. doi:10.1371/journal.pone.0076918.

Solowij, N., S. J. Broyd, C. Beale, J. A. Prick, L. M. Greenwood, H. van Hell, C. Suo, P. Galettis, N. Pai, S. Fu, R. J. Croft, J. H. Martin, and M. Yucel. 2018. "Therapeutic effects of prolonged cannabidiol treatment on psychological symptoms and cognitive function in regular cannabis users: A pragmatic open-label clinical trial." *Cannabis Cannabinoid Res* 3 (1): 21–34. doi:10.1089/can.2017.0043.

Solowij, N., S. Broyd, L. M. Greenwood, H. van Hell, D. Martelozzo, K. Rueb, J. Todd, Z. Liu, P. Galettis, J. Martin, R. Murray, A. Jones, P. T. Michie, and R. Croft. 2019. "A randomised controlled trial of vaporised delta(9)-tetrahydrocannabinol and cannabidiol alone and in combination in frequent and infrequent cannabis users: Acute intoxication effects." *Eur Arch Psychiatry Clin Neurosci* 269 (1): 17–35. doi:10.1007/s00406-019-00978-2.

Sonego, A. B., F. V. Gomes, E. A. Del Bel, and F. S. Guimaraes. 2016. "Cannabidiol attenuates haloperidol-induced catalepsy and c-fos protein expression in the dorsolateral striatum via 5-HT1A receptors in mice." *Behav Brain Res* 309: 22–28. doi:10.1016/j.bbr.2016.04.042.

Song, C., C. W. Stevenson, F. S. Guimaraes, and J. L. Lee. 2016. "Bidirectional effects of cannabidiol on contextual fear memory extinction." *Front Pharmacol* 7: 493. doi:10.3389/fphar.2016.00493.

Spindle, T. R., E. J. Cone, D. Kuntz, J. M. Mitchell, G. E. Bigelow, R. Flegel, and R. Vandrey. 2020. "Urinary pharmacokinetic profile of cannabinoids following administration of vaporized and oral cannabidiol and vaporized CBD-dominant cannabis." *J Anal Toxicol.* 44 (2): 109–125. doi:10.1093/jat/bkz080.

Spindle, T. R., M. O. Bonn-Miller, and R. Vandrey. 2019. "Changing landscape of cannabis: Novel products, formulations and methods of administraton." *Curr Opin Psychology* 30: 98–102. doi:10.1016/j.copsyc.2019.04.002.

Spinella, T. C., S. H. Stewart, J. Naugler, I. Yakovenko, and S. P. Barrett. 2021. "Evaluating cannabidiol (CBD) expectancy effects on acute stress and anxiety in healthy adults: A randomized crossover study." *Psychopharmacology (Berl).* 238 (7): 1965–1977. doi:10.1007/s00213-021-05823-w.

Srebnik, M., N. Lander, A. Breuer, and R Mechoulam. 1984. "Base catalyzed double bond isomerizations of cannabinoids: Structural and stereochemical aspects." *J Chem Soc., Perkin Trans* 1: 2881–2886.

Stark, T., J. Ruda-Kucerova, F. A. Iannotti, C. D'Addario, R. Di Marco, V. Pekarik, E. Drazanova, F. Piscitelli, M. Bari, Z. Babinska, G. Giurdanella, M. Di Bartolomeo, S. Salomone, A. Sulcova, M. Maccarrone, C. T. Wotjak, Z. Starcuk, Jr., F. Drago, R. Mechoulam, V. Di Marzo, and V. Micale. 2019. "Peripubertal cannabidiol treatment rescues behavioral and neurochemical abnormalities in the MAM model of schizophrenia." *Neuropharmacology* 146: 212–221. doi:10.1016/j.neuropharm.2018.11.035.

Stern, C. A. J., T. R. da Silva, A. M. Raymundi, C. P. de Souza, V. A. Hiroaki-Sato, L. Kato, F. S. Guimaraes, R. Andreatini, R. N. Takahashi, and L. J. Bertoglio. 2017. "Cannabidiol disrupts the consolidation of specific and generalized fear memories via dorsal hippocampus CB1 and CB2 receptors." *Neuropharmacology* 125: 220–230. doi:10.1016/j.neuropharm.2017.07.024.

Stern, C. A., L. Gazarini, R. N. Takahashi, F. S. Guimaraes, and L. J. Bertoglio. 2012. "On disruption of fear memory by reconsolidation blockade: Evidence from cannabidiol treatment." *Neuropsychopharmacology* 37 (9): 2132–2142. doi:10.1038/npp.2012.63.

Stern, C. A., L. Gazarini, A. C. Vanvossen, A. W. Zuardi, I. Galve-Roperh, F. S. Guimaraes, R. N. Takahashi, and L. J. Bertoglio. 2015. "Delta9-tetrahydrocannabinol alone and combined with cannabidiol mitigate fear memory through reconsolidation disruption." *Eur Neuropsychopharmacol* 25 (6): 958–965. doi:10.1016/j.euroneuro.2015.02.001.

Steru, L., R. Chermat, B. Thierry, and P. Simon. 1985. "The tail suspension test: A new method for screening antidepressants in mice." *Psychopharmacology (Berl)* 85 (3): 367–370. doi:10.1007/bf00428203.

Strumberg, D., S. Brugge, M. W. Korn, S. Koeppen, J. Ranft, G. Scheiber, C. Reiners, C. Mockel, S. Seeber, and M. E. Scheulen. 2002. "Evaluation of long-term toxicity in patients after cisplatin-based chemotherapy for non-seminomatous testicular cancer." *Ann Oncol* 13 (2): 229–236. doi:10.1093/annonc/mdf058.

Sultan, S. R., S. A. Millar, T. J. England, and S. E. O'Sullivan. 2017. "A systematic review and meta-analysis of the haemodynamic effects of cannabidiol." *Front Pharmacol* 8: 81. doi:10.3389/fphar.2017.00081.

Szabó, I. L., E. Lisztes, G. Béke, K. F. Tóth, R. Paus, A. Oláh, and T. Bíró. 2020. "The phytocannabinoid (-)-cannabidiol operates as a complex, differential modulator of human hair growth: Anti-inflammatory submicromolar versus hair growth inhibitory micromolar effects." *Journal of Investigative Dermatology* 140 (2): 484–488.e5. doi: 10.1016/j.jid.2019.07.690.

Szaflarski, J. P., E. M. Bebin, A. M. Comi, A. D. Patel, C. Joshi, D. Checketts, J. C. Beal, L. C. Laux, L. M. De Boer, M. H. Wong, M. Lopez, O. Devinsky, P. D. Lyons, P. P. Zentil, and R. Wechsler. 2018. "Long-term safety and treatment effects of cannabidiol in children and adults with treatment-resistant epilepsies: Expanded access program results." *Epilepsia* 59 (8): 1540–1548. doi:10.1111/epi.14477.

Sznitman, S. R., S. Vulfsons, D. Meiri, and G. Weinstein. 2020. "Medical cannabis and insomnia in older adults with chronic pain: A cross-sectional study." *BMJ Support Palliat Care* 10 (4): 415–420. doi:10.1136/bmjspcare-2019-001938.

Tagne, A. M., B. Pacchetti, M. Sodergren, M. Cosentino, and F. Marino. 2020. "Cannabidiol for viral diseases: Hype or hope." *Cannabis and Cannabinoid Research*. 5 (2): 121–131. doi:10.1089/cann.2019.0060.

Tahamtan, A., M. Tavakoli-Yaraki, T. P. Rygiel, T. Mokhtari-Azad, and V. Salimi. 2016. "Effects of cannabinoids and their receptors on viral infections." *J Med Virol* 88 (1): 1–12. doi:10.1002/jmv.24292.

Tamir, I., R. Mechoulam, and A. Y. Meyer. 1980. "Cannabidiol and phenytoin: A structural comparison." *J Med Chem* 23 (2): 220–223. doi:10.1021/jm00176a022.

Taylor, L., J. Crockett, B. Tayo, and G. Morrison. 2019. "A phase 1, open-label, parallel-group, single-dose trial of the pharmacokinetics and safety of cannabidiol (CBD) in subjects with mild to severe hepatic impairment." *J Clin Pharmacol* 59 (8): 1110–1119. doi:10.1002/jcph.1412.

Taylor, L., B. Gidal, G. Blakey, B. Tayo, and G. Morrison. 2018. "A phase I, randomized, double-blind, placebo-controlled, single ascending dose, multiple dose, and food effect trial of the safety, tolerability and pharmacokinetics of highly purified cannabidiol in healthy subjects." *CNS Drugs* 32 (11): 1053–1067. doi:10.1007/s40263-018-0578-5.

Tham, M., O. Yilmaz, M. Alaverdashvili, M. E. M. Kelly, E. M. Denovan-Wright, and R. B. Laprairie. 2019. "Allosteric and orthosteric pharmacology of cannabidiol and cannabidiol-dimethylheptyl at the type 1 and type 2 cannabinoid receptors." *Br J Pharmacol* 176 (10): 1455–1469. doi:10.1111/bph.14440.

Thiele, E. A., E. D. Marsh, J. A. French, M. Mazurkiewicz-Beldzinska, S. R. Benbadis, C. Joshi, P. D. Lyons, A. Taylor, C. Roberts, K. Sommerville, and Gwpcare Study Group. 2018. "Cannabidiol in patients with seizures associated with Lennox-Gastaut syndrome (GWPCARE4): A randomised, double-blind, placebo-controlled phase 3 trial." *Lancet* 391 (10125): 1085–1096. doi:10.1016/S0140-6736(18)30136-3.

Thomas, H. 1993. "Psychiatric symptoms in cannabis users." *Br J Psychiatry* 163:141–149. doi:10.1192/bjp.163.2.141.

Todd, S. M., and J. C. Arnold. 2016. "Neural correlates of interactions between cannabidiol and delta(9)-tetrahydrocannabinol in mice: Implications for medical cannabis." *Br J Pharmacol* 173 (1): 53–65. doi:10.1111/bph.13333.

Torres, S., M. Lorente, F. Rodriguez-Fornes, S. Hernandez-Tiedra, M. Salazar, E. Garcia-Taboada, J. Barcia, M. Guzman, and G. Velasco. 2011. "A combined preclinical therapy of cannabinoids and temozolomide against glioma." *Mol Cancer Ther* 10 (1): 90–103. doi:10.1158/1535-7163.MCT-10-0688.

Toth, C. C., N. M. Jedrzejewski, C. L. Ellis, and W. H. Frey, 2nd. 2010. "Cannabinoid-mediated modulation of neuropathic pain and microglial accumulation in a model of murine type I diabetic peripheral neuropathic pain." *Mol Pain* 6: 16. doi:10.1186/1744-8069-6-16.

Touitou, E., N. Dayan, L. Bergelson, B. Godin, and M. Eliaz. 2000. "Ethosomes—novel vesicular carriers for enhanced delivery: Characterization and skin penetration properties." *J Control Release* 65 (3): 403–418. doi:10.1016/s0168-3659(99)00222-9.

Tramer, M. R., D. Carroll, F. A. Campbell, D. J. Reynolds, R. A. Moore, and H. J. McQuay. 2001. "Cannabinoids for control of chemotherapy induced nausea and vomiting: Quantitative systematic review." *BMJ (Clinical Research Ed.)* 323 (7303): 16–21. doi:10.1136/bmj.323.7303.16.

Tran, T., and R. Kavuluru. 2020. "Social media surveillance for perceived therapeutic effects of cannabidiol (CBD) products." *Int J Drug Policy* 77: 102688. doi:10.1016/j.drugpo.2020.102688.

Trigo, J. M., D. Lagzdins, J. Rehm, P. Selby, I. Gamaleddin, B. Fischer, A. J. Barnes, M. A. Huestis, and B. Le Foll. 2016. "Effects of fixed or self-titrated dosages of Sativex on cannabis withdrawal and cravings." *Drug Alcohol Depend* 161: 298–306. doi:10.1016/j.drugalcdep.2016.02.020.

Trigo, J. M., A. Soliman, L. C. Quilty, B. Fischer, J. Rehm, P. Selby, A. J. Barnes, M. A. Huestis, T. P. George, D. L. Streiner, G. Staios, and B. Le Foll. 2018. "Nabiximols combined with motivational enhancement/cognitive behavioral therapy for the treatment of cannabis dependence: A pilot randomized clinical trial." *PLoS One* 13 (1): e0190768. doi:10.1371/journal.pone.0190768.

Trojano, M., and C. Vila. 2015. "Effectiveness and tolerability of THC/CBD oromucosal spray for multiple sclerosis spasticity in Italy: First data from a large observational study." *Eur Neurol* 74 (3–4): 178–185. doi:10.1159/000441619.

Turcott, J. G., M. Del Rocío Guillen Núñez, D. Flores-Estrada, L. F. Oñate-Ocaña, Z. L. Zatarain-Barrón, F. Barrón, and O. Arrieta. 2018. "The effect of nabilone on appetite, nutritional status, and quality of life in lung cancer patients: A randomized, double-blind clinical trial." *Support Care Cancer* 26 (9): 3029–3038. doi:10.1007/s00520-018-4154-9.

Turkanis, S. A., K. A. Smiley, H. K. Borys, D. M. Olsen, and R. Karler. 1979. "An electrophysiological analysis of the anticonvulsant action of cannabidiol on limbic seizures in conscious rats." *Epilepsia* 20 (4): 351–363. doi:10.1111/j.1528-1157.1979.tb04815.x.

Turri, M., F. Teatini, F. Donato, G. Zanette, V. Tugnoli, L. Deotto, B. Bonetti, and G. Squintani. 2018. "Pain modulation after oromucosal cannabinoid spray (*Sativex* in patients with multiple sclerosis: A study with quantitative sensory testing and laser-evoked potentials." *Medicines (Basel)* 5 (3): 59. doi:10.3390/medicines5030059.

Twomey, C. D. 2017. "Association of cannabis use with the development of elevated anxiety symptoms in the general population: A meta-analysis." *J Epidemiol Community Health* 71 (8): 811–816. doi:10.1136/jech-2016-208145.

Ueberall, M. A., U. Essner, and G. H. Mueller-Schwefe. 2019. "Effectiveness and tolerability of THC:CBD oromucosal spray as add-on measure in patients with severe chronic pain: Analysis of 12-week open-label real-world data provided by the German Pain e-Registry." *J Pain Res* 12: 1577–1604. doi:10.2147/JPR.S192174.

van Breemen, R. D., R. Muchiri, T. A. Bates, J. B. Weinstein, H. C. Leier, S. Farley, and F. G. Tafesse. 2022. "Cannabinoids block cellular entry of SARS-CoV-2 and the emerging variants." *J Natural Products*. Online ahead of print. doi:10.1021/acs.jnatprod.1c00946.

van de Donk, T., M. Niesters, M. A. Kowal, E. Olofsen, A. Dahan, and M. van Velzen. 2019. "An experimental randomized study on the analgesic effects of pharmaceutical-grade cannabis in chronic pain patients with fibromyalgia." *Pain* 160 (4): 860–869. doi:10.1097/j.pain.0000000000001464.

van Orten-Luiten, A. B., N. M. de Roos, S. Majait, B. J. M. Witteman, and R. F. Witkamp. 2021. "Effects of cannabidiol chewing gum on perceived pain and well-being of irritable bowel syndrome patients: A placebo-controlled crossover exploratory intervention study with symptom-driven dosing." Online ahead of print. *Cannabis Cannabinoid Res*. doi:10.1089/can.2020.0087.

Vann, R. E., T. F. Gamage, J. A. Warner, E. M. Marshall, N. L. Taylor, B. R. Martin, and J. L. Wiley. 2008. "Divergent effects of cannabidiol on the discriminative stimulus and place conditioning effects of Delta(9)-tetrahydrocannabinol." *Drug Alcohol Depend* 94 (1–3): 191–198. doi:10.1016/j.drugalcdep.2007.11.017.

Varvel, S. A., J. L. Wiley, R. Yang, D. T. Bridgen, K. Long, A. H. Lichtman, and B. R. Martin. 2006. "Interactions between THC and cannabidiol in mouse models of cannabinoid activity." *Psychopharmacology (Berl)* 186 (2): 226–234. doi:10.1007/s00213-006-0356-9.

Vaughn, D., J. Kulpa, and L. Paulionis. 2020. "Preliminary investigation of the safety of escalating cannabinoid doses in healthy dogs." *Frontiers in Veterinary Science* 7: 51. doi:10.3389/fvets.2020.00051.

Vaughn, D. M., L. J. Paulionis, and J. E. Kulpa. 2021. "Randomized, placebo-controlled, 28-day safety and pharmacokinetics evaluation of repeated oral cannabidiol administration in healthy dogs." *American Journal of Veterinary Research.* 82 (5): 405–416. doi:10.2460/ajvr.82.5.405.

Vecchio, D., C. Varrasi, E. Virgilio, A. Spagarino, P. Naldi, and R. Cantello. 2020. "Cannabinoids in multiple sclerosis: A neurophysiological analysis." *Acta Neurol Scand* 142 (4): 333–338. doi:10.1111/ane.13313.

Vela, J., L. Dreyer, K. K. Petersen, A. N. Lars, K. S. Duch, and S. Kristensen. 2021. "Cannabidiol treatment in hand osteoarthritis and psoriatic arthritis: A randomized, double-blind placebo-controlled trial." Online. *Pain.* doi:10.1097/j.pain.0000000000002466.

Vermersch, P., and M. Trojano. 2016. "Tetrahydrocannabinol:cannabidiol oromucosal spray for multiple sclerosis–related resistant spasticity in daily practice." *Eur Neurol* 76 (5–6): 216–226. doi:10.1159/000449413.

Vigil, J. M., S. S. Stith, I. M. Adams, and A. P. Reeve. 2017. "Associations between medical cannabis and prescription opioid use in chronic pain patients: A preliminary cohort study." *PLoS One* 12 (11): e0187795. doi:10.1371/journal.pone.0187795.

Villares, J. 2007. "Chronic use of marijuana decreases cannabinoid receptor binding and mRNA expression in the human brain." *Neuroscience* 145 (1): 323–334. doi:10.1016/j.neuroscience.2006.11.012.

Viudez-Martinez, A., M. S. Garcia-Gutierrez, A. I. Fraguas-Sanchez, A. I. Torres-Suarez, and J. Manzanares. 2018. "Effects of cannabidiol plus naltrexone on motivation and ethanol consumption." *Br J Pharmacol* 175 (16): 3369–3378. doi:10.1111/bph.14380.

Viudez-Martinez, A., M. S. Garcia-Gutierrez, C. M. Navarron, M. I. Morales-Calero, F. Navarrete, A. I. Torres-Suarez, and J. Manzanares. 2018. "Cannabidiol reduces ethanol consumption, motivation and relapse in mice." *Addict Biol* 23 (1): 154–164. doi:10.1111/adb.12495.

Volicer, L., M. Stelly, J. Morris, J. McLaughlin, and B. J. Volicer. 1997. "Effects of dronabinol on anorexia and disturbed behavior in patients with Alzheimer's disease." *Int J Geriatr Psychiatry* 12 (9): 913–919.

Vuolo, F., F. Petronilho, B. Sonai, C. Ritter, J. E. Hallak, A. W. Zuardi, J. A. Crippa, and F. Dal-Pizzol. 2015. "Evaluation of serum cytokines levels and the role of cannabidiol treatment in animal model of asthma." *Mediators Inflamm* 2015: 538670. doi:10.1155/2015/538670.

Wade, D. T., P. Makela, P. Robson, H. House, and C. Bateman. 2004. "Do cannabis-based medicinal extracts have general or specific effects on symptoms in multiple sclerosis? A double-blind, randomized, placebo-controlled study on 160 patients." *Mult Scler* 10 (4): 434–441. doi:10.1191/1352458504ms1082oa.

Wade, D. T., P. Robson, H. House, P. Makela, and J. Aram. 2003. "A preliminary controlled study to determine whether whole-plant cannabis extracts can improve intractable neurogenic symptoms." *Clin Rehabil* 17 (1): 21–29. doi:10.1191/0269215503cr581oa.

Wallace, D., A. L. Martin, and B. Park. 2007. "Cannabinoid hyperemesis: Marijuana puts patients in hot water." *Australasian Psychiatry* 15 (2): 156–158. doi:10.1080/10398560701196778.

Walsh, Z., R. Callaway, L. Belle-Isle, R. Capler, R. Kay, P. Lucas, and S. Holtzman. 2013. "Cannabis for therapeutic purposes: Patient characteristics, access, and reasons for use." *Int J Drug Policy* 24 (6): 511–516. doi:10.1016/j.drugpo.2013.08.010.

Wang, B., A. Kovalchuk, D. Li, R. Rodriguez-Juarez, Y. Ilnytskyy, I. Kovalchuk, and O. Kovalchuk. 2020. "In search of preventive strategies: novel high-CBD *Cannabis sativa* extracts modulate ACE2 expression in COVID-19 gateway tissues." *Aging (Albany NY)* 12 (22): 22425–22444. doi:10.18632/aging.202225.

Wang, Y., P. Mukhopadhyay, Z. Cao, H. Wang, D. Feng, G. Hasko, R. Mechoulam, B. Gao, and P. Pacher. 2017. "Cannabidiol attenuates alcohol-induced liver steatosis, metabolic dysregulation, inflammation and neutrophil-mediated injury." *Sci Rep* 7 (1): 12064. doi:10.1038/s41598-017-10924-8.

Wanner, N. M., M. Colwell, C. Drown, and C. Faulk. 2021. "Developmental cannabidiol exposure increases anxiety and modifies genome-wide brain DNA methylation in adult female mice." *Clinical Epigenetics* 13 (1): 4. doi:10.1186/s13148-020-00993-4.

Ward, S. J., S. D. McAllister, R. Kawamura, R. Murase, H. Neelakantan, and E. A. Walker. 2014. "Cannabidiol inhibits paclitaxel-induced neuropathic pain through 5-HT(1A) receptors without diminishing nervous system function or chemotherapy efficacy." *Br J Pharmacol* 171 (3): 636–645. doi:10.1111/bph.12439.

Ward, S. J., M. D. Ramirez, H. Neelakantan, and E. A. Walker. 2011. "Cannabidiol prevents the development of cold and mechanical allodynia in paclitaxel-treated female C57Bl6 mice." *Anesth Analg* 113 (4): 947–950. doi:10.1213/ANE.0b013e3182283486.

Watt, G., and T. Karl. 2017. "In vivo evidence for therapeutic properties of cannabidiol (CBD) for Alzheimer's disease." *Front Pharmacol* 8: 20. doi:10.3389/fphar.2017.00020.

Wei, D., H. Wang, J. Yang, Z. Dai, R. Yang, S. Meng, Y. Li, and X. Lin. 2020. "Effects of O-1602 and CBD on TNBS-induced colonic disturbances." *Neurogastroenterol Motil* 32 (3): e13756. doi:10.1111/nmo.13756.

Weiss, L., M. Zeira, S. Reich, M. Har-Noy, R. Mechoulam, S. Slavin, and R. Gallily. 2006. "Cannabidiol lowers incidence of diabetes in non-obese diabetic mice." *Autoimmunity* 39 (2): 143–151. doi:10.1080/08916930500356674.

Weiss, L., M. Zeira, S. Reich, S. Slavin, I. Raz, R. Mechoulam, and R. Gallily. 2008. "Cannabidiol arrests onset of autoimmune diabetes in NOD mice." *Neuropharmacology* 54 (1): 244–249. doi:10.1016/j.neuropharm.2007.06.029.

Welburn, P. J., G. A. Starmer, G. B. Chesher, and D. M. Jackson. 1976. "Effect of cannabinoids on the abdominal constriction response in mice: Within cannabinoid interactions." *Psychopharmacologia* 46 (1): 83–85. doi:10.1007/bf00421553.

Welch, S. P. 1997. "Characterization of anandamide-induced tolerance: comparison to delta 9-THC-induced interactions with dynorphinergic systems." *Drug Alcohol Depend* 45 (1–2): 39–45. doi:10.1016/s0376-8716(97)01342-2.

Wheeler, M., J. W. Merten, B. T. Gordon, and H. Hamadi. 2020. "CBD (cannabidiol) product attitudes, knowledge, and use among young adults." *Subst Use Misuse* 55 (7): 1138–1145. doi:10.1080/10826084.2020.1729201.

Whiting, P. F., R. F. Wolff, S. Deshpande, M. Di Nisio, S. Duffy, A. V. Hernandez, J. C. Keurentjes, S. Lang, K. Misso, S. Ryder, S. Schmidlkofer, M. Westwood, and J. Kleijnen. 2015. "Cannabinoids for medical use: A systematic review and meta-analysis." *JAMA* 313 (24): 2456–2473. doi:10.1001/jama.2015.6358.

Wiley, J. L., J. J. Burston, D. C. Leggett, O. O. Alekseeva, R. K. Razdan, A. Mahadevan, and B. R. Martin. 2005. "CB1 cannabinoid receptor-mediated modulation of food intake in mice." *Br J Pharmacol* 145 (3): 293–300. doi:10.1038/sj.bjp.0706157.

Wilkie, G., B. Sakr, and T. Rizack. 2016. "Medical marijuana use in oncology: A review." *JAMA Oncol* 2 (5): 670–675. doi:10.1001/jamaoncol.2016.0155.

Wilkinson, J. D., B. J. Whalley, D. Baker, G. Pryce, A. Constanti, S. Gibbons, and E. M. Williamson. 2003. "Medicinal cannabis: Is delta9-tetrahydrocannabinol necessary for all its effects?" *J Pharm Pharmacol* 55 (12): 1687–1694. doi:10.1211/0022357022304.

Wilkinson, J. D., and E. M. Williamson. 2007. "Cannabinoids inhibit human keratinocyte proliferation through a non-CB1/CB2 mechanism and have a potential therapeutic value in the treatment of psoriasis." *J Dermatol Sci* 45 (2): 87–92. doi:10.1016/j.jdermsci.2006.10.009.

Williams, C. M., N. A. Jones, and B. J. Whalley. 2014. "Cannabis and epilepsy." In *Handbook of Cannabis*, edited by Roger G. Pertwee, 547–564. Oxford: Oxford University Press.

Wilson, M. M., C. Philpot, and J. E. Morley. 2007. "Anorexia of aging in long term care: Is dronabinol an effective appetite stimulant? A pilot study." *J Nutr Health Aging* 11 (2): 195–198.

Withey, S. L., B. D. Kangas, S. Charles, A. B. Gumbert, J. E. Eisold, S. R. George, J. Bergman, and B. K. Madras. 2021. "Effects of daily delta(9)-tetrahydrocannabinol (THC) alone or combined with cannabidiol (CBD) on cognition-based behavior and activity in adolescent nonhuman primates." *Drug Alcohol Depend* 221: 108629. doi:10.1016/j.drugalcdep.2021.108629.

Wray, L., C. Stott, N. Jones, and S. Wright. 2017. "Cannabidiol does not convert to delta(9)-tetrahydrocannabinol in an in vivo animal model." *Cannabis Cannabinoid Res* 2 (1): 282–287. doi:10.1089/can.2017.0032.

Wright, M. J., Jr., S. A. Vandewater, and M. A. Taffe. 2013. "Cannabidiol attenuates deficits of visuospatial associative memory induced by delta(9) tetrahydrocannabinol." *Br J Pharmacol* 170 (7): 1365–1373. doi:10.1111/bph.12199.

Xu, D. H., B. D. Cullen, M. Tang, and Y. Fang. 2020. "The effectiveness of topical cannabidiol oil in symptomatic relief of peripheral neuropathy of the lower extremities." *Curr Pharm Biotechnol* 21 (5): 390–402. doi:10.2174/1389201020666191202111534.

Xu, X., M. Han, T. Li, W. Sun, D. Wang, B. Fu, Y. Zhou, X. Zheng, Y. Yang, X. Li, X. Zhang, A. Pan, and H. Wei. 2020. "Effective treatment of severe COVID-19 patients with tocilizumab." *Proceedings of the National Academy of Sciences* 117 (20): 10970–10975. doi:10.1073/pnas.2005615117.

Yang, K. H., S. Galadari, D. Isaev, G. Petroianu, T. S. Shippenberg, and M. Oz. 2010. "The nonpsychoactive cannabinoid cannabidiol inhibits 5-hydroxytryptamine3A receptor-mediated currents in *Xenopus laevis* oocytes." *J Pharmacol Exp Ther* 333 (2): 547–554. doi:10.1124/jpet.109.162594.

Yang, L., R. Rozenfeld, D. Wu, L. A. Devi, Z. Zhang, and A. Cederbaum. 2014. "Cannabidiol protects liver from binge alcohol-induced steatosis by mechanisms including inhibition of oxidative stress and increase in autophagy." *Free Radic Biol Med* 68: 260–267. doi:10.1016/j.freeradbiomed.2013.12.026.

Yeshurun, M., O. Shpilberg, C. Herscovici, L. Shargian, J. Dreyer, A. Peck, M. Israeli, M. Levy-Assaraf, T. Gruenewald, R. Mechoulam, P. Raanani, and R. Ram. 2015. "Cannabidiol for the prevention of graft-versus-host-disease after allogeneic hematopoietic cell transplantation: Results of a phase II study." *Biol Blood Marrow Transplant* 21 (10): 1770–1775. doi:10.1016/j.bbmt.2015.05.018.

Yimam, M., A. O'Neal, T. Horm, P. Jiao, M. Hong, S. Rossiter, L. Brownell, and Q. Jia. 2021. "Antinociceptive and anti-inflammatory properties of cannabidiol alone and in combination with standardized bioflavonoid composition." *J Med Food* 24 (9): 960–967. doi:10.1089/jmf.2020.0178.

Yucel, M., V. Lorenzetti, C. Suo, A. Zalesky, A. Fornito, M. J. Takagi, D. I. Lubman, and N. Solowij. 2016. "Hippocampal harms, protection and recovery following regular cannabis use." *Transl Psychiatry* 6: e710. doi:10.1038/tp.2015.201.

Zajicek, J., P. Fox, H. Sanders, D. Wright, J. Vickery, A. Nunn, and A. Thompson. 2003. "Cannabinoids for treatment of spasticity and other symptoms related to multiple sclerosis (CAMS study): Multicentre randomised placebo-controlled trial." *Lancet* 362 (9395): 1517–1526. doi:10.1016/s0140-6736(03)14738-1.

Zanelati, T. V., C. Biojone, F. A. Moreira, F. S. Guimaraes, and S. R. Joca. 2010. "Antidepressant-like effects of cannabidiol in mice: Possible involvement of 5-HT1A receptors." *Br J Pharmacol* 159 (1): 122–128. doi:10.1111/j.1476-5381.2009.00521.x.

Zhu, Y. F., K. Linher-Melville, M. J. Niazmand, M. Sharma, A. Shahid, K. L. Zhu, N. Parzei, J. Sidhu, C. Haj, R. Mechoulam, and G. Singh. 2020. "An evaluation of the anti-hyperalgesic effects of cannabidiolic acid-methyl ester in a preclinical model of peripheral neuropathic pain." *Br J Pharmacol.* 177 (12): 2712–2725. doi:10.1111/bph.14997.

Zuardi, A. W., R. A. Cosme, F. G. Graeff, and F. S. Guimaraes. 1993. "Effects of ipsapirone and cannabidiol on human experimental anxiety." *J Psychopharmacol* 7 (1 Suppl.): 82–88. doi:10.1177/026988119300700112.

Zuardi, A., J. Crippa, S. Dursun, S. Morais, J. Vilela, R. Sanches, and J. Hallak. 2010. "Cannabidiol was ineffective for manic episode of bipolar affective disorder." *J Psychopharmacol* 24 (1): 135–137. doi:10.1177/0269881108096521.

Zuardi, A. W., J. A. Crippa, J. E. Hallak, F. A. Moreira, and F. S. Guimaraes. 2006. "Cannabidiol, a *Cannabis sativa* constituent, as an antipsychotic drug." *Braz J Med Biol Res* 39 (4): 421–429. doi:10.1590/s0100-879x2006000400001.

Zuardi, A. W., J. E. Hallak, and J. A. Crippa. 2012. "Interaction between cannabidiol (CBD) and (9)-tetrahydrocannabinol (THC): influence of administration interval and dose ratio between the cannabinoids." *Psychopharmacology (Berl)* 219 (1): 247–249. doi:10.1007/s00213-011-2495-x.

Zuardi, A. W., S. L. Morais, F. S. Guimaraes, and R. Mechoulam. 1995. "Antipsychotic effect of cannabidiol." *J Clin Psychiatry* 56 (10): 485–486.

Zuardi, A. W., J. A. Rodrigues, and J. M. Cunha. 1991. "Effects of cannabidiol in animal models predictive of antipsychotic activity." *Psychopharmacology (Berl)* 104 (2): 260–264. doi:10.1007/bf02244189.

Zuardi, A. W., N. P. Rodrigues, A. L. Silva, S. A. Bernardo, J. E. C. Hallak, F. S. Guimaraes, and J. A. S. Crippa. 2017. "Inverted U-shaped dose-response curve of the anxiolytic effect of cannabidiol during public speaking in real life." *Front Pharmacol* 8: 259. doi:10.3389/fphar.2017.00259.

Zuardi, A. W., I. Shirakawa, E. Finkelfarb, and I. G. Karniol. 1982. "Action of cannabidiol on the anxiety and other effects produced by delta 9-THC in normal subjects." *Psychopharmacology (Berl)* 76 (3): 245–250. doi:10.1007/BF00432554.

# Index

Note: Page numbers in italics refer to figures.

(+) cannabidiol, 17

2-AG, 127, 137
4′-F-CBD, 18
5-HT receptors, 99–105, 113
5-HT$_{1A}$ receptor
  anxiety and, 154
  attenuated learned fear expression and, 165
  CBD and, 32, 36
  depression and, 170
  hypoxic ischemic brain injury and, 135
  ischemic events and, 70
  nausea and vomiting and, 100, 101, 103–104, 113
  pain and inflammation and, 121, 125–126, 128
  psychosis and, 185
  schizophrenia and, 191–192
5-HT$_3$ receptors, 98, 99–100
$\Delta^6$ isomer of CBD, 15
7-OH-CBD, 26
$\Delta^8$-iso-THC, 13–14
8-OH-DPAT, 100, 103

Abnormal CBD (abn-CBD), 13
Acetaminophen, 118
Acetic acid-induced pain, 122–123
Acetylcholine, 18, 175
Acidic conditions, chemical reactivity and, 13–14
Acne/acne vulgaris, 81, 85–87, 94, 219
Acoustic startle response, 184
Acute inflammation, 122–123
Acute nausea and vomiting, 98–99, 103–104, 114
Acute pain models, 120–122
Adams, Roger, 2, 12
Addiction, 195–205, 217–218
Addiction Research Center Inventory, 46
Adenosine, 18, 32
Agriculture Improvement Act (US; 2018), 7
Alcohol, 199–201, 204–205, 218
Allodynia, 118, 123, 125, 127, 128, 145, 148
Allogeneic hematopoietic cell transplantation (alloHCT), 74

Alprazolam, 151
Alzheimer's disease (AD), 79–80, 81, 215, 218
American Psychiatric Association, 181
Amisulpride, 187, 189–190, 210
Amygdala, 79, 160, 177, 191
Amyloid-β protein, 79, 215
Anandamide, 31, 40, 86, 121, 131, 137, 187–188, 192, 199
Angiogenesis, 91
Anhedonia, 169, 182, 195
Animal models of inflammatory conditions, 134–139
Anterior cingulate cortex, 129, 160
Anticipatory nausea and vomiting, 98–99, 104–105, 114, 217
Antiemetic and antinausea effects, 101–105. *See also* Nausea and vomiting
Antinociceptive effects, 43
Antioxidants, 16, 63–64, 70
Antipsychotic effects, 49–50
Anxiety, 47–48, 149–161, 209–210, 212
Apomorphine, 184
Apoptosis, 65, 69, 90
Appetite modulation, 108–112, 114–115
Arthritis, 145, 148, 213
Astrocytes, 64
Atopic dermatitis, 81, 82
Attenuated learned fear expression, 165–166
Autoimmune diseases, 72–76, 94, 215–216
Autoimmune encephalitis (AEA), 78–79
Autoimmune myocarditis, 75–76
Autophagy, 201

B cells (bone marrow- or bursa-derived cells), 65
Barbiturates, 152

Basal lipogenesis, 86
Basic conditions, chemical reactivity and, 15
Benzodiazepines, 149, 151, 152
Bioavailability, 21–26, 36
Bipolar disorder, 181, 190
Blood pressure, 33, 45
Bodily symptoms scale (bss), 155
Bone marrow transplants, 74–75
Breast cancer, 89–90, 216, 218
Brief Assessment of Cognition in Schizophrenia (BACS), 188
Brief Psychiatric Rating Scale (BPRS scale), 187
Buspirone, 149, 154

Cachexia, 97, 108, 111
Caco-2 cells, 136–137
Canadian Pain Society, 139
Cancer, 31, 87–88, 89–95, 216, 218
Cancer pain, 140, 144
Cannabidiolic acid (CBDA), 17–18, *18*, 67, 100, 101–102, 104–105, 125, 130–131, 153, 213, 217
Cannabidivarin, 56
Cannabinoid hyperemesis syndrome, 99, 112–113, 115
Cannabis
 addiction and, 202–203, 205
 compounds in, 11
 historical usage of, 1
 legal status of, 7–9
 research on, 1–6
Cannabis Act (Canada), 8
Capsaicin cream, 113, 115
Cardiomyopathy, 71
Cardiovascular effects, 71
Carrageenan-induced inflammatory pain, 129–131
Catalepsy, 185
Cataplexy, 177
Caudate nucleus, 192

CB$_1$ receptor, 17, 30–31, 40, 91, 101, 121–122, 131, 137, 146, 154, 185
CB$_2$ receptor, 17, 30–31, 40, 89, 90–91, 122, 131, 135, 136
CBDA (cannabidiolic acid), 17–18, *18*, 100, 101–102, 104–105, 125, 130–131, 153, 213, 217
CBDA+, 110
CBD-dimethylheptyl (CBD-DM), 20
Cell-mediated immunity, 65
Cerebral ischemia, 34, 69–71, 213–214
Chemical aspects of cbd, 12–20
Chemical reactivity of cbd, 13–20, *14*
Chemotherapy-induced nausea and vomiting, 97–99, 105–107, 114, 115, 217
Chemotherapy-induced neuropathic pain model, 125–127, 147–148, 213
Chromatography, 11
Chronic pain, 111, 117, 139, 140, 142–143, 173, 176
Chronic pruritus, 81
Cisplatin, 88, 126, 216
Citalopram, 151, 166
Clinical Global Impressions Scale (CGI-I and CGI-S), 188
Clobazam, 27, 58
Cocaine, 203–204
Cognitive behavioral therapy (CBT), 149
Cognitive performance, 39–40, 42–43, 48–49
Colitis, 72
Collagen, 132
Colon cancer, 91
Community Assessment of Psychic Experiences (CAPE), 50
Complete Freund's adjuvant-induced inflammatory pain, 131–133
Conditioned gaping, 102–105
Conditioned place preference, 35, 196–197
Conservative treatment, 142–143

Controlled Substances Act (US), 6, 7, 82
Coronavirus disease 2019 (COVID-19), 66–69
Crohn's disease, 76, 111, 135, 138–139, 211
Crystalline structure of CBD, 12–13
Cue-induced craving, 198–199, 209
Cutaneous endocannabinoid system, 85–86
CYP2C19, 27, 151
CYP2C34A, 151
CYP2D6, 118
CYP3A4, 27, 118, 151
Cytochrome (CYP) enzymes, 33, 36–37, 42
Cytochrome P450 (CYP) enzymes, 26–27
Cytochrome P450 enzymes, 33
Cytochrome P450 isoenzymes, 118, 133, 151
Cytokine storms, 67–68
Cytokines, 65, 73, 117, 132

D$_2$ receptors, 182, 184–185, 191
Delayed nausea and vomiting, 98–99, 106
Department of Agriculture, US, 82–83
Dependence, 34
Dependence and abuse potential of CBD, 34–36
Depression, 17, 163, 169–172, 217
Derivatives of CBD, 18–20, *19*
Dermatitis, 219
Dexamethasone, 68
Diabetes type I, 71, 73–74, 213, 215–216
Diabetic neuropathic pain, 127–129
*Diagnostic and Statistical Manual of Mental Disorders, Fifth Edition (DSM-5)*, 181
Diazepam, 152
Dopamine, 18, 35, 43, 80, 175, 182, 184–185, 191, 215
Dopaminergic neurotransmission, 184
Dorsal raphe nucleus, 103, 175

Dorsal striatum, 185
Double-blind, randomized, controlled trials (RCTs), 6
Doxorubicin, 71
Dravet syndrome (DS), 2–3, 6, 24, 58–61, 112, 115, 208, 212
Dronabinol, 105, 111, 115
Drug Enforcement Administration, US (DEA), 7, 11, 82
Drug interactions, 33, 37, 58
Drug testing, CBD and, 29–30

E-cigarettes, 22
Eczema, 84–85, 94
Elevated plus maze (EPM), 151–152
Endocannabinoid system, 1–2
Enterochromaffin cells, 98
Entorhinal cortex, 201
Entourage effects, 12
Epilepsy
  clinical trials involving, 2–3, 6, 24, 27, 33–34
  description of, 53
  FDA approval and, 208
  historical aspects of CBD and, 54–61
ERK1-2 pathway, 192
Ethosomes, 83
Ethylenediaminetetraacetic acid, 136–137
Experimental autoimmune hepatitis, 75
Experimental autoimmune myocarditis (EAM), 75–76
Exposure therapy, 164–165
Extracellular vesicle (EV) inhibitor, 92

Farm Bill (US; 2018), 7, 82–83
Fatty acid amide hydrolase (FAAH), 31, 40, 121, 131, 188, 192, 199
FDA approval, 2, 4, 5, 6, 8
Fear extinction, 165–166, 168
Fear-conditioning paradigm, 164–165
Fentanyl, 119, 197–198

Fibromyalgia, 146–147
Fluoxetine, 151
Focal seizures, 53
Food effects, 25–26
Forced swim test (FST), 169–171
Formalin-induced pain, 122–123
FosB/ΔFosB expression, 184
Free radicals, 63–64
Functional magnetic resonance imaging (fMRI), 160

Gamma aminobutyric acid (GABA), 53
Gaoni, Y., 1
Gastrointestinal tract inflammation, 135–137
Gender differences, 17, 125
Generalized anxiety disorder, 149
Generalized seizures, 53
Glioblastoma cells, 92
Glioma, 90, 216, 218
Global Assessment of Functioning scale (GAF), 188
Glutamate, 53, 55–56, 69–70
Glycine receptor subtypes, 32
GPR55 receptors, 32, 55–56, 192
Graft versus host disease (GvHD), 74–75, 94, 211, 212
Gut permeability, 135–138, 211
GW Pharmaceuticals, 24, 28, 51–52

Hair growth, 88
Health Canada, 8
Heart dysfunction, 214, 216
Heart rate, 33, 35–36
Hemp, legal status of, 7, 82–83
Hepatitis, 67, 75, 215–216
Hepatosteaosis, 200–201
Heroin, 197–198, 209
Hippocampus, 171, 192, 201
HIV, 66
HU-580, 100, 104, 114, 153, 171, 174, 175, 217

HUF-101, 122, 131
Humoral immunity, 65
Huntington's disease (HD), 80–81, 111, 215
Hydroxylation, 26
Hyperalgesia, 118, 123, 125, 127
Hyperinflammation conditions, 68
Hyperlocomotion, 183, 185–186
Hyperphosphorylated tau protein, 79
Hypothalamus, 175
Hypothermia, 44
Hypoxic ischemic brain injury, 134–135, 148, 214

Identification of CBD, 2
IFN-γ, 66, 73, 136
IL-1, 134
IL-6, 66, 68
IL-12, 73
Immune system, 65–66
In vitro studies, role of, 4
In vivo preclinical studies, role of, 4–5
Inducible nitric oxide synthase (iNOS), 66
Infections, increase in, 69
Inflammation
  acute, 122–123
  animal models of, 134–139
  carrageenan-induced, 129–131
  Complete Freund's adjuvant-induced, 131–133
  conclusions regarding, 147–148
  gastrointestinal tract, 135–137
  hyperinflammation conditions, 68
  inflammatory pain models and, 129–133
  introduction to, 117–119
  liver, 72
  see also pain
Inflammatory cytokines, 20, 31
Inflammatory pain models, 129–133
Insomnia, 173

Interferon gamma (IFNγ), 117
Interleukins (IL), 117
Interoceptive insular cortex (IIC), 103–104
Intoxicating effects, subjective, 45–46
Intracellular neurofibrillary tangles, 79
Intracranial self-stimulation (ICSS), 34–35
Irradiation of CBD, 16–17
Irritable bowel disease (IBD), 135–137, 138–139, 211, 212
Islets-of-Langerhans beta cells, 73
Isomerization, 15

Keratinocytes, 84–85, 87, 94
Kidney function, 71–72, 126
KLS-13019, 125–126

Labeling, problems with, 4
Learned fear expression, 165
Learned helplessness model, 171
Legal status of CBD, 7–8, 82–83
Lennox-Gastaut syndrome (LGS), 6, 24, 58, 60–61, 112, 115, 208, 212
Leukemia, 90–91
Light-dark emergence test, 152–153
Limbic system, 160
Lipopolysaccharide (LPS), 66, 136, 137
Liver function, 23, 26–27, 33, 71–72, 133–134, 200–201, 204, 214, 218
Locomotor suppressant effects, 44
Lung cancer, 91
Lymphoma, 90–91

Maastricht Acute Stress Test, 157
Macrophages, 65
Major depressive disorder (MDD), 163, 169–172. *See also* Depression
MATRICS Consensus Cognitive Battery (MCCB), 189
Mechoulam, R., 1, 2, 12, 56
Medial prefrontal cortex, 184

## Index

Melanoma, 87–88, 216, 218
Memory, 39, 42–43, 48–49, 164–165, 190
Memory consolidation, 165
Mesocorticolimbic dopamine activity, 43
Mesolimbic dopamine (DA) system, 182, 191
Metabolism of CBD, 26–27
Methadone, 119
Methamphetamine, 203
Methylazoxymethanol acetate (MAM), 185–186
Microglial cell activation, 66
MK167, 86
MK-801, 183–184
Moclobemide, 149
Morphine, 120, 121, 145–146, 147, 212
Multiple sclerosis (ms), 51, 77–79, 81, 139, 141, 144, 148, 211, 215, 218
Myeloid cells, 65
Myeloid-derived suppressor cells (MDSCs), 75
Myeloperoxidase (MPO), 65, 66
Myocardial infarction, 71
Myocarditis, 75–76, 215–216

Nabilone, 105, 111
Nabiximols
  appetite modulation and, 111
  approvals for, 51
  cancer and, 92, 144
  cannabis use disorder and, 202–203, 205
  Huntington's disease (HD) and, 81
  multiple sclerosis (MS) and, 77–78, 139–140, 144
  nausea and vomiting and, 99, 106–107
  pain and, 139–140, 143–147, 148
Narcolepsy, 177, 178
National Academy of Sciences, 106, 111, 139, 154–155

Nausea and vomiting
  5-HT receptors and, 99–105
  chemotherapy- or radiation-induced, 97–99
  in humans, 105–108
  preclinical studies on, 216–217, 218
Nausea suppression, 17, 34, 42–43
N-desmethylclobazam, 27
Neurodegenerative diseases, 77–81, 94
Neurogenerative disorders, 215
Neuroimaging studies, 159–160
Neurokinin-1 (NK1) receptors, 98
Neuropathic pain, 17, 117–118, 124–129, 141, 145, 147–148, 212, 213
Neuropathic pain models, 123–129
Neutrophils, 65, 66
NF-kB activity, 20, 66, 71
Nicotine, 199, 204, 218
Nigrostriatal pathway, 184–185
Nitric oxide (NO) production, 31, 65–66
Nociceptive pain, 117–118
Nonrapid eye movement (NREM) sleep, 172–173
Nonsteroidal anti-inflammatory drugs, 118
Nuclear magnetic resonance (NMR) data, 12
Nucleus accumbens, 175, 191

Olanzepine, 98
Olivetol, 13
Ondansetron, 217
Opiate addiction, 197–199, 204, 209, 212
Opioids, 118–119, 139–140, 146, 147
Oral administration, 23–26
Osteoarthritis models, 133–134
Oxaliplatin, 126
Oxford Liverpool Inventory of Life Experiences, 50
Oxidation, 15–16, *15*
Oxidative stress, 64, 71–72, 79, 84–85, 89, 90, 91

Paced Auditory Serial Addition Task, 48–49
Paclitaxel, 125–126
Pain
  chronic, 111, 117, 139, 140, 142–143, 173, 176
  conclusions regarding, 147–148
  human studies and, 139–147
  human trials and, 211–212
  introduction to, 117–119
  nabiximols and, 143–147
  neuropathic, 123–129
  osteoarthritis models and, 133–134
  preclinical studies on, 119–123, 213
  *see also* inflammation
Paranoia, 181
Parkinson's disease (PD), 80, 159, 177–178, 209, 215
Permeation enhancement, 83
Peroxidase proliferator–activated receptor γ (PPARγ), 32
Pharmacodynamics of CBD, 30–32, 31$t$
Pharmacological aspects of CBD, 20–37
Phases of clinical trials, 5–6
Phenobarbital, 27
Phenyl benzoquinone-induced pain, 122–123
Phenytoin, 55
Photochemical reactions of CBD, 16–17, 16
Phytocannabinoids, 11
P-mentha-2,8-dien-1-ol, 13
Positive and Negative Syndrome Scale (PANSS scale), 187, 188, 189
Postoperative pain, 129, 145–146, 148, 213
Postpartum psychosis, 181
Posttraumatic stress disorder (PTSD), 149, 163–168, 177, 209–210, 217
P-quinone, 15–16
Pregabalin, 149
Pregnancy, 32

Prepulse inhibition (PPI), 183–184, 186
Psoriasis, 84–85, 94, 219
Psychiatric clinic patients, 159
Psychomotor performance, 46–47
Psychosis, 49–50, 159, 181–193, 209
Pulse rate, 44–45

Radiation-induced nausea and vomiting, 97–99
Randomized, controlled trials (RCTs)
  barriers to, 7–9
  in phase 3 trials, 6
Rapid eye movement (REM) sleep, 173, 177–178
Rapid treatment, 142–143
Reactive oxygen species (ROS) production, 31, 64, 70, 87, 90, 91, 93
Reconsolidation, 165, 168
REM sleep latency, 173–174
Remdesivir, 67
Rostral anterior cingulate cortex, 129
Routine treatment, 142–143

Safety and toxicology, 32–34
Safety and toxicology of CBD, 32–34
Schedule I substances, 7, 8
Schizophrenia, 50, 181–193, 210, 212
Schizotypal Personality Questionnaire, 50
Sciatic nerve model, 124–125, 213
Sebocytes, 86, 94
Sebum production, 85–86
Selective serotonin reuptake inhibitors (SSRIs), 149, 166
Self-administration procedure, 196, 197, 200, 218
Serotonin, 18, 32, 98, 128, 171, 175
Serotonin 3 (5-HT$_3$) receptors, 98, 99–100
Serotonin-norepinephrine reuptake inhibitors (SNRIs), 149
Severe acute respiratory syndrome coronavirus 2 (SARS-CoV-2), 66–69

Simulated public speaking (SPS) test, 44, 155–156, 158, 159
Skin cancer, 87–88, 216, 218
Skin disease, 81–88, 94
Sleep, 47, 172–178
Sleep disorders, 176–178
Sleep latency, 173–174
Smoking, 22
Social anxiety disorder, 149, 158, 160, 209
Social withdrawal, 183
Spasticity, 77–78, 141, 144, 148, 212
Spectrometry, 11
Spinal cord injury, 127, 148
Spontaneous pain, 127
Spontaneously hypertensive strain of rats (SHR), 186
Standardization, lack of, 4
State-trait anxiety inventory (STAI), 48, 155
Steady-state CBD plasma concentrations, 25
Stimulants, 203–204
Stress reduction, 17
Submucosal spray, 23
Substance p, 98
Substance use disorder, PTSD and, 166–167
Sweet solutions, liking of, 109–110, 114
Synovial membrane, 132–133
Synthesis of CBD, 13, *13*
Systemic lupus erythematosus (SLE), 76

T cells (thymus cells), 65, 66, 75–76
Tail suspension test (TST), 169–170
Temozolomide, 92
Terpenes, 11
$\Delta^9$-tetrahydrocannabinol (THC)
chemical reactivity and, 13–14
IBD and, 137
identification of, 1
in illicit products, 11–12
metabolism of, 27
schizophrenia and, 182
structure of, 13
$\Delta^8$-THC, 13–14
THC
anxiety and, 151, 155
concentration of, 11–12
conversion into, 28–29
interactions with, 39–52
schizophrenia and, 191–192
$\Delta^9$-THCOOH, 29–30
Thyroid cancer, 91
TMOG cells, 20
TNF production, 20
TNF-α, 73, 117, 132–133, 136
Tocilizumab, 68
Tocopherol, 90
Todd, A., 2, 12
Tolerance, 34
Topical application, studies on, 121, 140, 142, 145, 148, 212
Topical formulations, 22–23
Topoisomerase II inhibitors, 15–16
Transcription factors, 31
Transdermal delivery, 23, 83, 132, 142
Transient receptor potential (TFP) family, 32
Triazolam, 151
Tricyclic antidepressants, 149
TRPV1 receptors
anxiety and, 154
cancer and, 89, 91
cannabinoid hyperemesis syndrome and, 113, 115
epilepsy and, 55
hepatitis and, 75
pain and, 124, 130, 131, 133, 136
TRPV4, 86
Tuberous sclerosis complex, 6, 208, 212

Ulcerative colitis, 72, 135–136
Urine drug testing, 29–30
UV radiation, 87

Vanilloid receptor type1 (VR1), 192
Vaporizing, 22, 28
Ventral tegmental area (VTA), 191
Verbal recall, 48
Vero cells, 67
Vincristine, 126–127
Viral diseases, 66–69
Visual analogue mood scale (VAMS), 155
Visual analogue scales (VAS), 45–46
Vitamin E acetate, 22
Vomiting. *See* Nausea and vomiting

WAY100635, 185
Whole plant cannabis, 99, 111